W9-CAF-128

The Death of RELIGION and the Rebirth of SPIRIT

The Death of RELIGION and the Rebirth of SPIRIT

A Return to the Intelligence of the Heart

JOSEPH CHILTON PEARCE

Park Street Press
Rochester, Vermont

Park Street Press
One Park Street
Rochester, Vermont 05767
www.InnerTraditions.com

Park Street Press is a division of Inner Traditions International

The Library of Congress Cataloging-in-Publication Data

Pearce, Joseph Chilton.
The death of religion and the rebirth of spirit : a return to the intelligence of the heart / Joseph Chilton Pearce.
p. cm.
Includes bibliographical references and index.
ISBN-13: 978-1-59477-171-2 (hardcover)
ISBN-10: 1-59477-171-5 (hardcover)
1. Spirituality—Psychology. 2. Love—Religious aspects. 3. Violence—Religious aspects. 4. Human evolution—Religious aspects. 5. Religion—Controversial literature. I. Title.
BL624.P425 2007
200—dc22

2006103206

Printed and bound in the United States by Lake Book Manufacturing

10 9 8 7 6 5 4 3 2 1

Text design and layout by Rachel Goldenberg
This book was typeset in Sabon, with Trajan as the display typeface.

To

Robert Sardello

CONTENTS

Part Three

THE REBIRTH OF SPIRIT AND THE RESUMPTION OF EVOLUTION

ACKNOWLEDGMENTS

My grateful thanks to editors Elaine Cissi and Vickie Trihy for their splendid work, making rough passages smooth, my crooked logic straight, and my wayward prose comprehensible. While pointing up contradictions, inconsistencies, comfortable prejudices, half-truths, and outright errors, they made it all challenging and fun, a playful dialogue of reciprocal elucidations—a writer could ask for no more.

Special thanks to J. W. Travis, M.D., who gave so generously of his time to help prepare the final edits of this book. Travis and his wife have formed the Alliance for Transforming the Lives of Children, a worldwide movement, and their *Wellness Workbook,* encompassing the best of common sense and sane medical advice, is deservedly a classic.

Robert Sardello was a major influence on the final shape of this book, which I had thought finished when someone sent me Sardello's *Love and the World.* I have borrowed liberally from this extraordinary work, which, along with his recently published *Silence,* has seriously influenced my way of thinking and personal life.

As I did for my book *The Biology of Transcendence,* I have borrowed heavily from the Institute of HeartMath, and I am grateful to them.

My very recent discoveries of David Loye and Riane Eisler forced additional last-minute changes in this book, and to their insights and knowledge I am grateful, both personally and on behalf of this book.

Several people struggled through early rough drafts of this work:

Bill and Win Sweet, Bob and Kathy Simmons, Robert Sardello, Michael Mendizza, and Jon Graham, of Inner Traditions • Bear & Company, are among them. Their encouragement in spite of the vagueness and occasional vapid nature of early drafts was encouraging. My long-suffering wife put up with my disappearance into the computer for long periods (some three years) during this writing, read the earlier drafts of the work, and offered practical suggestions for making it more readable and reasonable.

I have avoided those nuisance footnotes, op. cit., and et al. clutters that generally attend scholarly or academic efforts and toward which neither this book nor I have pretensions. The bibliography lists works particularly influential in this writing, however, whether or not I mention these in the text.

Please note that throughout the book and particularly in part 2, I use the word *him* for infant/child and the word *her* for mother. I tried *him/her* and *he/she* in my earlier writing to avoid any chauvinism, and it was a pain for both myself and the reader.

Part One

Culture as a Negative Field Effect and the Phenomenon of Mind

INTRODUCTION TO PART ONE

Cultures have risen and fallen throughout history, and when they fall, it has always been by their own hand. Whether or not by our own hand, our culture is rapidly waning as a widespread anxiety waxes. Philosopher Susanne Langer claimed that our greatest fear is a "collapse into chaos should our ideation fail," and culture is a major plank in our ideation. Threaten our fabric of beliefs, practices, and perspectives that make up the system of our cultural ideation and our very sense of self is threatened.

As culture is a major plank in our ideation, religion is a major plank in our culture, and it, too, is on the wane, which has given rise to fundamentalism as a political-cultural force. Arising from the adherents to all religious systems, old and new, fundamentalists fuel the fire of the very cultural collapse we fear.

At the same time, our current scientific technologies, which have become an even more powerful plank in our culture's ideation, damage us on every hand—physically, mentally, and morally—and because their work is indirect and subtle, it goes unrecognized. As we used to turn to religion as our hope and solace, we now turn to science, a religion with its own brand of protective fundamentalism. Both of these religions, scientific and ecclesiastic, are equally destructive to spirit, mind, and nature and equally give rise to violence and civility's decline.

While we do have a culture, our actions are hardly civil, and in spite of our many religions, a spiritual void seems epidemic. The mounting tide of violence toward self, earth, and others intensifies, while sporadic movements toward a spiritual renewal fragment in uncertainty. The impending death of religion, however, could bring—or at least allow—the rebirth of spirit.

In his book *The Ascent of Humanity,* Charles Eisenstein graphically shows that a compulsion within us for prediction and control is a fundamental flaw in our human venture. Every plank in the cultural ideation handed down to us rests on prediction and control—of ourselves, each other, nature, and the world. We assume that such a complex system of compulsive drives is instinctive and natural and we can't really think outside the boundaries of prediction and control even to question that which seems illogical. Our whole cultural fabric of mind arises from, sustains, and is sustained by this compulsion for prediction and control.

We think of prediction and control as high water marks of human intellect arising from a basic instinct to survive. But this compulsion is itself the force that leads to our demise, though stating it so baldly makes it incomprehensible to our cultural mind-set. In fact, we automatically rationalize to protect ourselves against such a notion and tend to screen out works and writing that bring it to our attention.

In his books *Love and the World* and *Silence,* Robert Sardello has taken the work of the late Austrian philosopher-scientist Rudolf Steiner to a new level of evolutionary thought. Steiner, a Ph.D. from a German university, is one of the great and woefully neglected minds of recent centuries. That Steiner's astonishing output has remained largely obscure and unknown almost a century after his death is a cultural effect that shields from us that very opening of mind toward which he himself pointed. Yet Steiner's lack of recognition and acceptance holds a key to our dilemma, for he leads beyond that cosmology of prediction-control that makes up the very warp and woof of contemporary thought.

A strikingly similar case in point is the history and scope of Charles Darwin as revealed in David Loye's recent book *Darwin's Lost Theory of Love.* That David Loye's work on Darwin has also been largely ignored, in spite of the acceptance and acclaim of Loye's earlier works, strangely parallels the history of Darwin's own work as well as that of Rudolf Steiner.

In his book, Loye describes how Darwin, in the latter part of his life, went beyond the accepted thought of his time to explore biology's relevance not only to a theory of evolution but also to what we know today as the fields of psychology, anthropology, brain science, and moral philosophy. *The Descent of Man,* Darwin's final work (which I will hereafter refer to as Darwin 2), is distinct from, yet complementary to, his earlier and widely accepted study, *The Origin of Species* (which I will refer to hereafter as Darwin 1).

Both are works of brilliance and insight, though Darwin 2 had a markedly different reception than did Darwin 1. Loye examines the strange fact that the last work of Darwin has been ignored, while his first has long been an accepted part of modern academic and scientific thought. In this contrast lies not just the evolutionary history of humankind but also an explanation for why we tremble at the gates of disaster today.

In Darwin 1, Darwin clearly describes how ages of mutation, selectivity, and survival of the fittest gave rise to mammalian life in general. On this foundation, as Darwin 2 shows, evolution then employed markedly different forces—"higher agencies," as Darwin called them—to bring about the far more advanced human species.

Using copious quotes from Darwin, Loye shows that these "higher agencies" translate as love of both self and other. Certainly, terms such as *love* and *altruism* are hardly in keeping with current academic and scientific (neo-Darwinian) acceptance as critical forces in evolution. Yet if love and altruism were developed, they would be the basis for not just our survival but also our recovery of an ongoing evolutionary momentum we have lost.

The catch lies in the word *developed.* If we look at contemporary societies worldwide and our historical record in general, we find a marked failure to develop love and altruism, even though remarkable forces apparently gave rise to us. I propose here that the higher force of love of self and other is both our true nature and the substantive foundation of our genetic system as described in Darwin 1. Further, I suggest that this higher force moves as powerfully as ever in us today, for our stake in evolution is evolution's stake in us, a typical strange-loop phenomenon.

These higher agencies are a combination of an instinctual base of love expanded upon by—and functioning through—nurturing. Nurturing goes far beyond simply nursing infants, as our simian ancestors did. According to Darwin 2, benevolent instincts of nurturing and care were the evolutionary springboard for our appearance, which may have been more recent than we have considered up to now. Recently, geneticists at the Howard Hughes Medical Center traced the DNA records of the major species preceding us and made the audacious claim that the human brain's evolutionary appearance was far too sudden to be accounted for by Darwin's selective mutation and survival of the fittest. But their discovery fits well with the Darwin 2 thesis explored by David Loye: The human brain to which these geneticists refer is the fourth and last in a long line of evolutionary neural systems carried within our skulls and, up until this latest one, developed over the ages through methods described in Darwin 1. Once this critical foundation of ancient Darwin 1 systems was completed and the supporting cast for us newcomers on life's stage had been well selected and rehearsed, nature could add the final fourth brain.

During many decades as head of the National Institute of Health's Department of Brain Evolution and Behavior, neuroscientist Paul MacLean mapped out the evolutionary nature and structure of our brain. He clearly showed that in our head lies the Darwin 1 foundation of an ancient reptilian or hind brain, which served as the basis for a forebrain consisting of an old and new mammalian brain. Upon these three evolved structures, the fourth human brain could be added with little of the slow, trial-and-error processes of the evolution leading up to it. This Darwin 2 phenomenon of a human brain operating from the higher agencies of love and altruism apparently brought us about as recently as forty to fifty thousand years ago, merely yesterday on the evolutionary timeline. Standing squarely on the shoulders of eons of Darwin 1 process, we with our fourth brain are apparently quite new on the neural scene.

Now, as our history and present circumstances indicate, and as this book will explain, our newest brain is continually being dominated and overruled by those very ancient systems on which it rests—systems that function largely through instinct rather than intelligence. Despite what we would expect from evolution's design, our history illustrates a constant struggle between, rather than synchrony of, the old and new neural systems in our head. A severe imbalance between defensive, old-brain instincts and intellectual new-brain systems is evidenced in our continual outbursts of violence and destruction.

These periodic seizures of violence and destruction seem to be not so much old-brain upheavals—those ancient systems aren't smart enough to engineer the fiendish means by which we kill each other—as upheavals of our new brain caught up in or seduced by the instinctual drives of the three older brains. This periodic seduction of the new by the old is in opposition to an overall evolutionary drift and totally counter to the higher agencies that brought us about. This continual usurping of the capacities of the new brain by the old has resulted in the fact that our new, fourth brain is largely undeveloped. Driving us to predict and control a nature and world we then can't trust, these upheavals either indicate a breakdown in evolution's biological plan or show that the plan is not yet complete and nature is still working out the glitches, searching for some design in which evolution can continue instead of self-destructing.

Actually, the neatly linear order of appearance in the evolutionary unfolding of the neural systems in our head may be misleading. Certainly, our oldest reptilian brain came first in evolution and gave rise to what followed. In fact, this earliest neural structure comes first in fetal brain growth and is first to develop after birth, followed in both cases by the old and new mammalian brains. The interplay of the three systems paves the way for the fourth human brain, which builds its structure after birth. In our notion of evolution-as-progress, then, we answer our compulsion for prediction and control. We assume evolution produced our advanced intelligence to predict and control earlier and inferior forces. Yet this neatly linear progression is a deceptive half-truth, for it overlooks and betrays the principle part of the creative process that underlies evolution itself and gave rise to us.

Consider turning upside down this notion of moving from inferior to ever more superior forms, for this conventional, common-sense notion puts the horse before the cart and its driver whereas with us humans, the driver came before the horse. The more advanced an evolutionary neural system, the more fragile it is. Our Darwin 2 brain, with its much higher form of intelligence, is radically dependent on these earlier neural systems for its own functioning, so the most logical and perhaps only feasible method of progression would have been for this higher Darwin 2 intelligence first to work out, through the slow and careful Darwin 1 selective process, what the higher had to have as its foundation in order to be.

A favorite quote of mine that neatly encapsulates this procedure is from Meister Eckhart: "There is no Being except through a Mode of Being." The "Being" here is our unknown and unknowable creator, life itself, and we are its mode, its means for being. William Blake said, "God only Acts and Is, In existing beings or Men." Consider, then, that the new Darwin 2 system, by which "Being itself" actually could be, was the initial impetus for the evolution of its forerunner.

Here, then, is an example of a strange-loop phenomenon: A new potential, sensed within an evolutionary process moving infinitely in all directions, brought about the appearance of what seems to be an older system required by the newer one. This strange loop is a major factor in creation and evolution found throughout the world and probably the cosmos.

Our failure to recognize this strange loop constitutes a lapse in our current knowledge and understanding, although such an interdependence has been recognized by other cultures and civilizations before ours. Neo-Darwinism, a limited and fragmented scientific view of Darwin 1, has vainly sought to prove that a "mode of being" gave rise to Being itself, which is patent nonsense.

Ironically, were the higher Darwin 2 forces of nurturing fully developed, such "superstructural drives," as Loye calls them, would, in times of stress and crisis, prove to be far more powerful and efficient than the foundational survival instincts nested in our primary reptilian brain so cherished by neo-Darwinist scientific and social disciplines. Our new human brain can simply outperform those ancient instinctual survival brains by a huge, incalculable margin. That we have not developed these higher systems and the lifelong mutual nurturing and altruism indicated in Darwin 2, and that we are subject instead to some pretty stupid moves prompted by our lower, instinctual systems is starkly evident today. In actuality, we use this incredible new brain on behalf of fear-driven survival instincts arising from that oldest evolutionary brain, which is a seriously devolutionary move that keeps us subject to instinct and compromises our intelligence.

Both our current religions, scientific and ecclesiastic, may well be offended at my contention here that they are destructive to life and civilization, that they are not nurturing but are, in fact, devolutionary. Yet recognition of this devolutionary effect is necessary if we are to clear the decks and open ourselves again to the evolutionary force of love and altruism that seems to lie behind our life and cosmos. These higher intelligences, giving rise to us, are our true spirituality and would be served by a true science.

In *Ascent of Humanity,* Charles Eisenstein puts forth his "separation-unification" theory to show how our separation of mind and heart was brought about by our compulsion for prediction and control and how unification depends on dropping this ancient habitual drive—admittedly, no easy matter. Nature did not evolve humanity that this humanity might turn around and attempt to predict and control nature's infinitely open system of balances. We are ourselves in and of nature; the balance lies within us first and foremost, and the work of Rudolf Steiner and Robert Sardello, as well as the invaluable research at the Institute of HeartMath, can consciously lead us to the heart, the source of that primal being of love and altruism. It is the heart, after all, that can lead us beyond all need for prediction and control, thus making our destructive compulsions obsolete.

As for my claim that religion and technological science are destructive, the religious community would answer that only a few bad apples have caused trouble in the past. Religious leaders might caution against throwing their baby out with the bath water. For their part, the technological-scientific community points out that the fault lies with the politicians, military leaders, corporate powers, and the like who use and abuse their gifts; it believes itself, the sacrosanct scientists, and their methods to be above reproach. Throwing out its baby is even more foolish, it points out, while the bath water should be bottled as though it were from Lourdes.

In science and technology we have created a self-propelling machine we can't turn off, however, and like the sorcerer's apprentice, we are overwhelmed by forces unleashed through our arrogance and ignorance. Every brilliant solution our technology and science have thus far presented has set up a counterwave of quiet, subtle, slow, and patient destruction, just as religions rapidly give rise to noisy and turbulent violence.

It is interesting to note that scientists such as Steiner and Darwin achieved their great works without employing technology in our modern sense. Theirs was the power of the human mind turning within to its own process, not turning without through artifice. "No one can know the joy I experience in just thinking," commented Steiner. Darwin achieved his first great insights through nontechnological observations of the living world as it is (or was), and those of his second, more mature stage of thought were arrived at through gardening, beekeeping, and acute awareness of life in his native English countryside. These are real-world processes hardly found in the explosive proliferation of destructive virtual realities so often relied upon by contemporary science. It is instructive to consider that the scientific insights of Darwin and Steiner did not lead to the damage and destruction of humanity and earth everywhere visible today.

In trying to cope with the hydra-headed assaults on humanity and nature wrought by both religion and technological science, we lose all trace of the origin of these assaults; we become so caught up in dealing with their harmful effects that we can't see their causes. Cultural anthropologist Leslie White observed that a culture self-destructs when the problems it produces outstrip its capacity for solution. When every move we make seems flawed by hidden error and every correction of an error creates two more errors in its wake—as seems the case today, no matter how sophisticated and scientific our apologetic terminology or how lofty and pompous our religious moral protest—the ground beneath us simply crumbles.

Over a half century ago philosopher Susanne Langer made the observation that we would do well to reconsider our unquestioned belief that modern science is a blessing to humanity. This was the middle of the twentieth century, the age of science coming into its own, with endless wonders and powers holding all in their thrall, like a new religion. Even as a casual aside, Langer's observation was as rank a heresy as possible in her day, as it probably would be in ours.

Envision a late medieval philosopher-critic suggesting, at the peak of cathedral building, that we might do well to reconsider our unquestioned belief in God. A true believer, facing Chartres Cathedral, one of the most beautiful and perfect structures conceived by man, would ask how we might be so blind as not to see the handiwork of God etched into every stone. In the same way, as we behold the modern world of virtual reality and the daily appearance of new miracles, wonderworks beyond the grasp of the ordinary mind, inventions that blind us by the light of our own brilliance that created them, we might ask how a philosopher—and woman at that—could question the ultimate goodness of all this largesse.

Susanne Langer's mentor, mathematician-philosopher Alfred North Whitehead, once proposed that science and technology could have arisen only in a Christian culture, though their roots are even more ancient. Examine only one thread in the rich sequence behind such a notion: Following the Greek influence on the inventive creations of Paul the Apostle's Christology, Christianity had demonized body on behalf of soul, declared a state of war between spirit and flesh, and pronounced nature the archenemy to be vanquished, brought to her knees, and made to yield her secrets and do our bidding. As a result of this conquest, a scientific priesthood arose that overshadowed its waning ecclesiastic parentage. Ultimately, in richest irony, the priests of each faith, old religion and new science, have played mock battle before a hapless humankind that has lost out all the way around.

For generations we were led to believe we had to choose between science and religion, which often seemed like a choice between being hanged or shot. But science and religion have not staked out all the territory available to our mind. In truth, we don't have to buy into either of these camps, nor will their proposed truce and merger prove the panacea we have long sought. Mating two mongrels doesn't produce a thoroughbred. There are countless other ways life can be lived.

Years ago, David Bohm, Einstein's protégé and physicist at the University of London's Birkbeck College, wrote of the substrate of reality being an "implicate order" of energy that is consciousness itself. This is a turnabout of conventional scientific theory, which assumes consciousness arose from aggregates of matter. Yet while Bohm's theory proposing that matter arises from aggregates of conscious energy seemed altogether new, it was an observation made by Shaivite scholars in Kashmir, India, ten centuries ago. Poet William Blake observed that spirit creating matter was a wondrous miracle, while the notion that matter could create spirit was sheer lunacy.

By an implicate order David Bohm meant a single underlying energy that has implied within it all potential, all possible fields of energy. A field is a particular aggregate or grouping of energy that arises from and gives rise to a particular aspect of our reality. An explicate order makes explicit or tangible some aspect of that implicate order of all conceivable potential. In just this way, we can see how the implicate potentials of a higher mind would express themselves through a lower system that the higher mind required in order to become explicit—as we find in the theory that the Darwin 2 system gave rise to Darwin 1. You can't have implicate without explicate, just as explicate relies on implicate. Their existence is interdependent and neither ever proves conclusively to take precedent over the other.

Rupert Sheldrake, a biologist from Cambridge University, speaks for a new science breaking out of the restriction of mind that academic science has long imposed. In the mid-1980s Sheldrake and Bohm, both scientists with strong spiritual foundations, held a series of dialogues on consciousness as the underlying substrate of our reality, the field of all fields, and field effect as the shaping force in all aspects of our life—in fact, in the entire working of our cosmology. Because we are part and parcel of that very consciousness, exploring this field effect can take us beyond the narrow constrictions of both science and religion and open us to the full dimensions of mind in creation or mind as creation—not that we might play God with a free hand but simply to stop our demonic self-destruction. What happens then, as evolution picks up where it left off and moves us on, is an unknown.

In part 1 we will explore both culture as a major implicate force shaping our explicate life and field effect as a shaping force in culture. The nature of field effect in general and the nature of mind, the recipient of these unseen forces, will be our focal point. Through examples, we'll clarify such terms as *field effect* and *mind.* The mirroring relationships or strange loops of cause and effect, field and mind, question and answer, discovery and creation, in which each seems to give rise to the other and the very existence of each relies on the other, as explored by cognitive scientist Douglas Hoffstadter in his book *Godel, Escher, and Bach,* will be examined here.

Within our mind is a neutral ground between the closed boundaries of science and religion. It is a ground explored by Robert Sardello and before him by Rudolf Steiner, who, of all scientists, recognized the key role the heart plays in this neutral ground that reveals itself as a heart-brain dialogue with no boundaries or binding principles, only its nonjudging creative force. This heart-mind interplay of consciousness is found not in an examination or analysis of our past or any combination of past notions, but only where, as Sardello richly puts it, the "future flows into the present." A rare unity of mind, heart, and spirit can open us to this neutral ground of mind and its endless strange loops that Rudolf Steiner points to in both his works *Knowledge of the Higher Worlds and How to Attain Them,* and his elusive thoughts in his book *Approaching the Mystery of Golgotha.*

You may or may not recall or have even heard of that lonely hill called Golgotha, the "place of the skull" on which stood, as our poet laureate Howard Nemerov said:

> *. . . The sticks and yardarms of the holy three-*
> *Masted vessel whereon the Son of Man*
> *Hung between thieves . . .*

There was burned into our collective psyche an event that cracked our cosmic egg. And though we have since sealed and resealed that crack again and again, it is always opening for us as—and if—we choose to open to it.

1

CULTURE AND DARKNESS OF MIND

Prisons are built with stones of law, brothels with bricks of religion.

WILLIAM BLAKE, *The Marriage of Heaven and Hell*

If we wish to study a particular organism closely and without interference in a laboratory project, we make a culture of the creature in an artificial environment we can control, such as a Petri dish or test tube. In this closed world we provide sufficient nutrients to keep the creature alive while isolating it from nature or the world at large, with all its random-chance side effects and unwanted influences that can clutter the scene. Thus we maintain control of the experiment while subtly entering into and changing the overall nature of what we're studying.

Human culture, as the term is used in this book, is a manmade function not so much distinct from nature as isolated from it. As theologian James Carse points out in his book *Finite and Infinite Games,* although everything, from first to last, is an aspect of nature, some aspects more than others are more cluttered with useless and harmful human debris. Culture as a habitual mental process has grown out of attempts to control nature and predict its events—even as culture is, by default, a factor in the nature we examine and attempt to control. In fact, our own human nature, with its wild assortment of behaviors, is the principle target of control. But we are culture itself, and we attempt to control our actions through the abstract force of a culture we create through such actions and are, in turn, subject to. A strange loop indeed.

Charles Darwin proposed that any activity repeated long enough will become a habit, and any habit repeated long enough can become instinctual, operating automatically below reason or the limen of our awareness. Culture is actually a collection of habits repeated over millennia, and many of these habits have become so instinctual as to function automatically, beneath our awareness. These are imprints that form our own consciousness, to varying extent, and that are passed on to our children both deliberately and unconsciously.

PASSING CULTURAL PATTERNS TO THE NEXT GENERATION

A new discovery in neuroscience concerns *mirror neurons,* large groupings of cells scattered throughout our brain that, beneath our awareness, automatically mirror or imprint various aspects of the world around us, locking them into our memory and cognitive process. Patricia Greenfield, neuroscientist at UCLA, claims: "Mirror neurons provide a powerful biological foundation for the evolution of culture." Previously, she observes, scholars have treated culture as fundamentally separate from biology, "but now we see that mirror neurons absorb culture directly, with each generation teaching the next by social sharing, imitation and observation."

"Social emotions like guilt, shame, pride, embarrassment, disgust and lust are based on a uniquely human mirror neuron system found in a part of the brain called the insula," reports Christian Keysers, who studies the neural basis of empathy at the University of Groningen in the Netherlands. While mirror neurons aren't exclusively human but are also found in many higher mammals, they play a major role in how we humans pass cultural patterns to the next generation without being aware of it.

One ordinary, commonsense view of culture recognizes it as a means (found even in higher primates) of passing on to offspring any acquired knowledge of world and self. A large percentage of our brain develops after birth and, as suggested by the theory of mirror neurons, is profoundly subject to such influences from parents, mentors and caretakers. So, while instinctual capacities for growth and response are passed on through genes, more advanced evolutionary creatures also pass on to their young their acquired, learned capacities for dealing with the world—critical adaptive skills.

Somewhere in our past, these cultural patterns became not just imitative but enforced behavioral modifications based on fear of both the natural world and our mentors and models who insisted we conform. These models were likewise culturally formed as infants, as were those who came before them for millennia. Eventually, group safety became associated with predictable actions of individual members, giving rise to a rich fabric of taboo, law, prohibitions, and conventions, along with punishments and reprisals for failure to conform. Attempts to predict and control extended to nature herself, and an ever more complex system of taboo and conventions arose, leading to our most recent science and technology.

In all this process, we've overlooked a problem brought about by this compulsion for conformity: The change our culture enforces and engenders alters the characteristics of the nature with which we begin. The very precautions and modifications of our offspring's behavior—those "gifts" that supposedly help our children to survive nature or the world—in fact change their own personal nature and, eventually, our mutual, shared natures and world. This changing nature means that generation by generation we up the ante on the extent to which we must enforce such modifications, creating a loop of cause and effect, with each bringing about the other.

CONFORMING: THE ROLE OF MONOCULTURE IN RISING VIOLENCE

Thus we are subject to both influences from our parents, mentors, and caretakers and instinctual genetic responses. Group safety is associated with predictable actions of members of the group, and miscreants must be weeded out. Cultural patterns become linked with survival instincts based on both fear of the natural world and the insistence of our mentors, models, and other social members that we conform. The grim injunction "Do this or else!" has been goading us for ages. In this way, we have unconsciously created a Petri dish for our minds, a mental test tube into which we have tried to stuff life itself in our attempts to ward off the random-chance aspects of creation and each other and to predict and control both nature and ourselves. Our efforts at control ironically add more unpredictable randomness and often upset whatever natural balance we begin with, making it necessary for us to employ even more stringent efforts at control.

These attempts to predict and control social behavior eventually extended to nature itself, that infinitely variable system of checks and balances, and our built-up cultural legacy became an infinite regress. Today, culture has become a primary formative force, a major organizing field of energy distinct from nature, which works to isolate us from what was, is, and might be natural. We might look on culture as a surrogate parent acting somewhat in the manner of our real mother, nature, but on behalf of an arbitrary system foreign to and against our real heritage. We are like children kidnapped by a foreign power and brought up to serve nefarious schemes.

Under the impact of these cultural schemes as they evolved, social consciousness and a benevolent civility seem to fade, even as indications from archaeology, anthropology, genetics, and pre- and perinatal research indicate that we are born with a natural benevolent response to each other, a built-in love of life, self, and others. As suggested in the introduction to part 1, the birth of the human brain-mind could not be accounted for through the same instinctual drives giving rise to species.

Giving rise to us and our fundamental nature was a new form of consciousness defined as altruism and love.

We have been separated from this original humanity by culture, which has itself become an automatic reflex, a near-unconscious, animal instinct. In fact, to think or act outside cultural dictates, once imprinted, has become almost impossible. This cultural effect has expanded to such power that it has become the very fabric of our conscious awareness—an awareness that results from being conceived, born, and brought up in it. Taking over and shaping our brain-minds accordingly, culture shapes even our attempts to examine or become objectively aware of its very force.

What has resulted is a single monoculture sweeping the globe and bringing a mounting tide of irrational and ever more intense violence. A faulty assumption is that this increasing violence results from a clash of cultures worldwide (from dumb foreigners who think differently than we do) and that when a single mind-set, even now apparently coming about, finally takes over and all subcultures are destroyed, the violence will subside.

But violence, as the French historian René Girard and his protégé Gil Bailie point out, is inherent within culture. In fact, the two, culture and violence, bring each other into being. What's more, a "global mind" already exists, to our detriment, in the form of the very substrate of culture that shapes our lives and now floods the planet with a uniform field of electronic media, with its integral structure of violence. Beneath the surface froth of colorations; quaint novelties of local effects; and different languages, rituals, myths, beliefs, and pathologies found in the many varieties of world culture lies the single substrate of culture as a force field, rather like one of Carl Jung's archetypes. It is this force that accounts for an overall mass conformity of mind beneath the various surface colorations.

Cultural conformity compromises individuality, a mind that can think outside the constraints of either our animal heritage or our culture and its global effects. Only an individual mind can pick up on the evolutionary drive of life itself and create independent of the artificial overlays and restrictions culture imposes. We humans were made for and long for that lost individuality.

The Force of Spirit-as-Life and the Force of Cultural Violence

Culture as a disruption in evolution, an issue addressed in the mystery of Golgotha (as we shall see in part 3), gave and still gives rise to religion, one of culture's primary foundations and means for sustenance. This point would hardly be acceptable to Girard or Bailie, but I stand with poet William Blake who claimed there is no natural religion. Religion did not arise from spirit, to which it is antithetical, but, ironically, it is sustained by that longing that it seems to engender in us. Thus spiritually starved, we turn to culture's religious counterfeit as the way out.

In fact, I use the words *spirit* and *spiritual* with misgivings because they have been so tarnished by hucksters, but as it is used here, spirit is life itself longing for expression. The Neoplatonic philosopher Plotinus spoke of the world as love expressing itself and spirit is that mode of being to itself that springs forth as life. Spirit-as-life loves itself because it is love itself. Only as we love ourselves passionately do we love life. Life cries out exultantly, "lift up the stone and I am there, break the stick and I am there." (Howard Nemerov, several times our nation's poet laureate and a self-described Jewish agnostic, says, "I split the stick . . . [and] saw nothing that was not wood, nothing that was not God.") Self-love is the foundation of spirit, just as loathing of self is the basis of religion and the culture that spawns it. The branding of self-love, the matrix of all love, as a pathological narcissism is a crime against our nature and a typical cultural ploy that ends in destruction of self and its larger body of earth.

In his book *Nature's Destiny,* New Zealand biochemist Michael Denton claims that the universe, from its first instant of existence, has moved inexorably to produce life. Denton's proposal harkens to the notion of *teleology,* a belief in the universe's primal purpose or goal, a notion that has long been a prime heresy in the culturally subservient academic sciences. Thus, in deference to current custom, Denton acknowledges that while the universe moves through random chance, this method of creation is stochastic in its unfolding. *Stochasm* is a Greek term referring to randomness with purpose—and just as an origin out of random chance is an evil notion to religion, the notion of purpose is anathema in scientism, as is anything smacking of teleology as related to the universe's end goal or that toward which evolution moves.

Here we explore how life-as-spirit is a force that culture usurps, leaving, according to academic belief, only a mechanical or chemical reaction or function, which should therefore be predictable and controllable by scientific method. Of course, it is also then devoid of recognition or awareness of spirit, leaving scientism as the only god around. Spirit is that which "bloweth as it listeth and no man knows its coming and going"—a random and unpredictable whimsy that just won't do for the mentality of control that indirectly drives both religion and scientism.

Culture and its religion and scientism would contemptuously deny this theft of spirit as they substitute an endless expanse of counterfeit trivia and magical manipulations to sustain their control over our behavior. In the irony of the ages, we are caught up in our attempts to ameliorate the remaining hunger of life that cannot be assuaged by anything other than the very spirit that is life itself, that embraces everything, including the very air we breathe. (Bernadette Roberts, mystic, teacher, and one-time Carmelite nun who spoke of having been lifted up into a state "absolutely other" to her ordinary self, breathed a "divine air" for some three weeks before our cultural world reasserted itself, the air polluted, the divine upstaged, the world too much with us.)

Admittedly, I have somewhat personified or made an object of culture, which is actually a psychological or mental field effect, a sphere of influence or force like gravity, intangible except for the disastrous fallout it produces. Culture as a field effect can contain within it a multitude of subcultures, which, in their energy, novelty, and markedly different textures, colors, and languages, are constantly in conflict with each other. With violence and religion as both the products and producers of culture, the resulting tangled web often obscures the presence of culture itself.

How does culture engender violence? It is brought on by the constant pressure of restrictions, prohibitions, and ceaseless demands for conformity to its abstractions. These constraints block our longing for life, which, ironically, intensifies accordingly. Of incidental consequence is that individual or group toward which cultural violence is directed. We always seem to find or create a target for our violence, whether internal or external. Should our internal violence leading to illness and neuroses fail to fill the bill, ideological clashes between subcultures are always with us, providing obvious external targets.

American culture produces a rich variety of internal projections of its violence, punctuated with periodic and necessary external violence in the form of war. War is that organized, legitimized, religiously sanctioned, and ritualized murder that boils up and expresses the rage arising from creatures cut off from nature and spirit and locked into subservience to culture. Our accumulated violence gives rise to an undercurrent of rage and hatred. We may keep it well-cloaked by a thin covering of civility, but we periodically act out these accumulated impulses, which are then always expressed as a well-rationalized necessity. As Robert Sardello tells us, the atomic bomb, created by an effort rationalized as absolutely necessary to survival, represents the final epitome of hatred, even as those scientists preparing it won Nobel prizes for their efforts.

Interestingly, the United States has, not only the largest collection of atomic bombs, the ultimate symbol of cultural hatred, but also the largest prison population of any nation. More subtle imprisonment can be found in our continued subjugation of women, which, though less overt than stoning them to death, as is still sanctioned beneath the auspices of the Muslim religion, nonetheless is ill treatment that serves our corporate-economic machine and harms all of us, males included.

And the violence extends to self: In myriad subtle but effective ways, we drive our children to internal or self-violence, resulting in one of the largest populations of unhealthy children in the world and the highest level of child suicide of any nation. Directly or indirectly, we sell those children the guns to slaughter each other in the more dramatic and publicized external forms of youthful rage.

Interestingly enough, media largely ignores child suicide and children's violence against each other unless these are spectacularly sensational—all this while we as a culture are urged to glorify and sanctify our heroes killed in various wars. Fifty thousand of our young people under age twenty-five were killed by each other in the United States between the years 1990 and 2000, about the same number of young men killed in our abortive, irrational war in Vietnam, deaths that earned each a place on that memorial wall in our nation's capital. Where is the memorial for these fifty-thousand youngsters? State-sanctioned murder, or those murdered in carrying out these sanctions, is always lauded by our war-makers. The fact that this time-worn cultural ploy began unraveling in that Vietnam travesty, near unraveling our country in the process, brought a note of hope not yet extinguished.

THE ROLE OF LAW IN CULTURE

While a principle characteristic of culture is religion and its ever-waxing violence, a second, far more pervasive, and ever-expanding element is law. Law, made necessary by violence, is itself a cause and a form of violence. Law is our god, just as our religious God is a God of law.

We have untold thousands of laws on the books, requiring legions of lawyers (whose numbers increase exponentially each year) to counsel us on protecting ourselves from that law or from other lawyers as neighbor battles neighbor in courts while battalions of jurists keep score and hand out verdicts proclaiming winners and losers. Much of this keeps the public enthralled and the media busy, while a nested hierarchy of judges mete out punishments and rewards according to law, precedence, evidence, or, in some cases, seeming personal whim. (Interestingly enough, the cultural symbol of the law is a blindfolded female in flowing gown holding a balance—indicating impartiality and adherence to "facts." But the symbol reads another way: Not only must she be blindfolded to take part in the process, she represents that hierarchy of judges and all beneath her who are blind to all human elements and are lost in the darkness of intellectual and cultural abstractions or political-economic gain.)

We keep our prisons filled with the indigent and poor who can't afford lawyers or who are assigned indifferent ones while the rich create their own laws that make them wealthier and protect their fortunes. This admired thievery has a long historical precedence in the millennia of tyrants, emperors, princes and kings, dictators, popes, and presidents who have fed on the social body while culture holds them up in the history books as models for our children.

Virtually every breath in our enculturated life is monitored by law, either blatantly or subtly. We live in a web of legal constraint and limitation accepted and adhered to unconsciously and unconditionally. From ancient commandments learned in Sunday school to the stately volumes of philosophies learned in higher education, we find an underlying strata of that cultural dictum: "Do this or else!" As a result, we live in the shadow of angst less we fail somehow and "they" find us guilty of some infringement of law—perhaps of one that we were not aware even existed. Taxes, speed limits, bills, enculturation of our children from birth, schooling, and college—all are part and parcel of the ever-expanding web of limitation and constraint that makes up life in culture to which we must adhere . . . or else!

One of the mythical convictions imprinted by culture is that without it and its laws, enforcements, and constraints, we would behave in a far more bestial manner than we do now. This "truism" ignores the fact that few animals (only certain chimpanzees under certain situations) will behave in so wantonly cruel a manner as we do under the existence of law. It seems clear, then, that our "bestial" behavior may well be a result of the limitations, constraints, and deliberate inhibiting interventions of culture itself. Of course, this is a wildly improbable notion to our enculturated minds, but one substantiated in this text. Because cultural constraints are unconscious behavioral patterns that shape our very mind and thought, we can't recognize them or their effects even as we are crippled by them.

One of the lies we are told as students is that history is a legitimate and necessary study, for through it we can see our past mistakes and profit from them—a nonsensical notion. We have never learned from history and apparently never will. While history may be largely fiction anyway, as historian Will Durante claimed decades ago, whatever its actual makeup, we simply repeat our past errors again and again, generally in ever new colors or camouflage. Our current actions (tomorrow's history), stirring up our adrenaline worldwide as never before, may well break these ancient cycles once and for all, for our mounting violence escalates with every effort we make to go beyond it. Yet we have within us a genetically encoded biology of transcendence or inherent ability and instinct to go beyond limitation and constraint, and this is our way out. Love and benevolence brought us into being, make up our natural constitution, and are ever ready to be reinstated.

In this chapter we have explored the general characteristics of culture as a field of influence. There is nothing academically quantifiable in this hypothesis of field effect; rather like gravity, there is no proof that it exists other than pointing to its results. Cultural effect is similar to movements within a collective unconscious as proposed by Carl Jung, an assumption that can open doors of the mind. A formative field of culture is a hypothesis through which we can view events and make sense of them and open again to our true nature.

2

CULTURE AND WAR

The tree of liberty must be periodically watered by the blood of patriots and tyrants.

THOMAS JEFFERSON

War is a direct cultural effect. Why do we engage in it again and again, generation after generation, millennia after millennia? We do so because war is absolutely necessary to culture, as Gil Bailie and René Girard claim, and we are blindly autonomous servants of the cultural effect, having been born and raised within it. As it is a sure bet that an individual will speak French if born and brought up in a French-speaking family and community, so will culture cast its far more powerful generic influence on those born to it. Culture as a field effect functions much like language acquisition: it results in a spontaneous, imitative learning below the limen of our awareness.

So what is it in culture that leads to an endless call-to-colors and wars mounted in fury and thoroughness, generation after generation? We killed some one hundred million of ourselves in the twentieth century through the wonders and efficiency of those most jealous cultural gods, science and technology, and because of their efficiency, we probably have but few wars left before we are altogether eliminated—humanity destroyed by its own cultural petard.

I enlisted as an aviation trainee in the Air Corps in the latter part of World War II not for the glory of flying and fighting for the Army Air corps, U.S. flag, or any other noble cause but to avoid being drafted as cannon fodder for the countless battalions of foot soldiers joined in mutual slaughter across the globe. The only alternative was prison, and partly because, as many teenagers do, I felt oddly invulnerable, I chose chancing bullets as the lesser evil.

The war ended before I finished training and could be sent overseas to possibly die (though surely as a hero), but nearly as many boys died in training as in aerial combat. In our nation's frantic push for replacements for that mayhem in the skies, some hundred thousand of us junior birdmen were crowded into the airways in flight training when the war's end was obviously in sight. As we advanced toward sky fighter overseas status, our training planes exploded, collided with each other, fell apart in midair, and piled into every kind of obstacle, killing us in countless ways before we could kill in even more countless ways in other skies.

But the Army Air Corps cadets had the best-looking uniforms, lived high on the hog as far as housing and food were concerned, and attracted the prettiest girls. Movies glamorized us, and many a romantic, sentimental song was crooned about us airmen off in the wild blue yonder with our silver wings. This was enough to attract any hot-blooded young man who aspired to far more than dying in the mud of Italy or the jungles of the South Pacific. Either way, we youngsters died by the millions while killing untold other millions.

SLAUGHTERHOUSE FIVE

At seventy-six years of age when the Twin Towers came down in gory 9/11 flame and everlasting fame, I glanced briefly at the televised footage of that spectacle and assiduously avoided looking at it again. Having no television in my house, I was to some extent spared direct exposure to the vast negative field that was building through our benevolent government's manipulations. I felt no shock at that report of those now-famous images, written of ad nauseam since—for a more powerful resonant memory overlay it all and I saw not so much the Twin Towers flaming but my long-ago inner image of the bombing of Dresden, Germany, in World War II. Played against the backdrop of that now-aging and near-forgotten war, 9/11 brought a single phrase to my mind, one that wouldn't stop repeating: "As you sow, so shall you reap." The phrase is hardly new, but it's meaning is true for individuals, nations, and the world.

Reports of our bombing of Dresden filtered through slowly. The allies were massively pounding Germany with retaliatory bombs of every description, blasting and firebombing cities to near oblivion (as "they" had assaulted "us"), when, in desperation, the German high command packed an estimated four hundred thousand women and children into Dresden, thinking that in all of Germany it was the one unlikely target for revenge. Dresden was an ancient medieval walled city where the only product for centuries had been fine china; it had no strategic value at all.

The allied high command, however, hearing of this massive refugee center, decided to destroy Dresden in order to break the spirit of the German people and thus supposedly shorten the war. Winston Churchill made the final decision, an act of cold hatred and culturally righteous revenge that was carried out by the British Royal Air Force. An estimated one hundred thousand women and children were cremated that first night, while those not blown to smithereens at the outset were torched by the firebombs. Death lingered long and hard behind those ancient walls.

Some ten years after that war, by chance my next-door neighbor was a young woman who had lived through the Dresden holocaust. She gave graphic descriptions of climbing down into the sewers of the city, where she and thousands of others spent four days in pitch darkness, linked arm in arm to form long human chains in order to hold each other against the tide of fetid water that flowed through, waiting for the flames and heat to subside. Invariably, someone would loose their grip and let go, with the chain regrouping to hold. I learned that what I thought a myth might well be true: people's hair can turn white overnight.

Germany's spirit was broken, I suppose, if any remained, and mine was peculiarly dampened on hearing and then reading the news reports, all revived years later by my neighbor's living and harrowing accounts. Dresden, however, had been but a dress rehearsal for the sons of liberty and justice.

The radio announcements of the atomic annihilation of Hiroshima and, shortly after, Nagasaki, where we incinerated unknown tens of thousands of Japanese women and children—admittedly, as their airmen would have incinerated us—came through quickly. The second bomb was loosed, it was later reported, to see what difference in effect there would be between a direct ground contact explosion of this new wonder of science and one exploded in the air. (Little was made of the fact that it was dropped on Nagasaki before the Japanese government had a chance to recover from the shock of Hiroshima and surrender.) It was a simple, neat, and quantifiable kind of scientific experiment. Indeed, Oppenheimer spoke of the first atomic explosion in Nevada as the most beautiful creation of mankind.

So to me, with World War II and its more than thirty million dead standing as a major milestone in my youth and seeming only yesterday, the Twin Towers episode, which killed three thousand people, was almost incidental. In the light of our onerous past, this was but a minor reaping of some major seed we had sown years ago. The mills of the gods of war grind as slowly as any other. (In the hysteria our nation's benevolent leaders worked up following 9/11, I almost expected arrest for my alien thoughts, treasonous emotional indifference, and refusal to be afraid of or take too seriously terrorists in general.)

Going back a few generations, before World War I, which had been over for eight years when I was born and of which I had no childhood impressions or memories, America killed 750,000 of its own men in the War between the States, or the Civil War, as we call it (though what could be civil about murdering 750,000?). Right here on our own soil there was a conflict that brought untold and ongoing suffering ignored by history books (which are so often written by the victors).

As a child I heard "grandmother tales" from both sides of my family in Virginia and South Carolina concerning that War of the Invasion, as they called it, when the northerners came to loot, rape, and burn far and wide. These tales were so vividly seared into my memory when I was a child that I considered that war and its aftermath as my war. Throughout my childhood we played Civil War, instead of the more recent World War I, which had no place at all in our mind's eye. Poison gas, machine guns, and tanks offered small competition to our soldier play of Yankees and Confederates, which was truly rivaled only by cowboys and Indians. But World War II made even these rich grandmother tales a minor footnote, until upstaged by 9/11.

SHOCK AND AFTERSHOCK

Surely, having madmen turn the tables of our own madness on us in this modern electronic age of control brought waves of shock and aftershock after 9/11. Yet periods of shock are times of great opportunity. The mind—even the mind of a nation—stops briefly at such times, and with the right leadership during that post-9/11 period we could have changed the course of history, for, because of the media, the shock resonated worldwide. We could have broken the ancient cycle of reciprocity, justice-seeking, revenge, and retaliation locked into the cultural mindset, the fuses that have ignited every war and changed the very shape and nature of a violence-prone species.

We could have initiated a transformative shift in the very mental apparatus that drives us to our self-destruction again and again. We had the opportunity, obscure as it now seems, to break the cycle of cultural enslavement, thereby freeing humanity from culture and dealing a death blow to that darkest and most powerful of all the principalities and powers to arise in our sad history. Many people were primed and ready for a radical shift, but we had no leaders at all who could envision this course of action—only a more lowly group of schemers than usual who in a base, despicable, and even demonic way used every event, calamity, and disaster to further their own greed and lust for power, all the while hiding behind the cross for camouflage, justification, and public support. (My personal resentment of the government's treachery lay more in their hiding behind the cross rather than the flag, since hiding behind the flag is a common political ploy, while the cross was and is an issue close to my heart, and all of which will be picked up again here in chapter 12 and part 3.)

A leader with the intentions of Gandhi might have had the personal stature, character, and almost superhuman courage and love of neighbor called for by 9/11 to break the vicious cycle of centuries of enculturation. But our so-called leaders were far from the models of both Gandhi and Jesus (to whom they referred time and again). And, as it turned out, neither Gandhi nor Jesus were successful anyway. Gandhi had moved against blatant cultural oppression, of course, and may to some extent have shifted the outer trappings of his culture of the moment, but his life and actions, unwittingly and indirectly, worked for the maintenance of culture as a reciprocal process or force. Any energy, whether positive or negative, expended toward culture strengthens culture. Jesus may have had in mind the erasure of culture as a force—but his earlier attempt met with no more success than Gandhi's later one, and for the same cultural reasons: Culture as a force simply destroyed him, as it had Gandhi, and assumed exploiting both of these figures to its own advantage. We still may hope, however, that the better part of the recollections concerning Jesus, particularly those concerning the graphic nature of his destruction, have left its own form of psychic imprint, which may yet achieve his goal.

CULTURE AS A FIELD EFFECT

What exactly do I mean by *culture,* that construct that I obviously use as a whipping post? And what is the field effect that gives culture its apparent permanence and power, driving nation to murder nation in chronic warfare? Here we look more closely at the axiom that our personal, social, and species-wide curse is culture.

If culture is a shaping force of probably unknowable origin, any attempt to explain or define the word is handicapped, for by default we have all been enculturated and to objectively examine the effect is akin to getting out of our own skin. An enculturated mind *is* culture, and the force of culture is directly dependent on our mind responding according to our own enculturation. In implicating culture as our hubris and nemesis, we implicate ourselves, an uncomfortable and threatening direction to pursue, one that immediately puts us on guard, for we automatically deflect any direct negative reference to our own personal sense of being. In this we seek to preserve our own integrity, a reasonable survival instinct and effective defense maneuver that operates largely beneath our awareness and often drives us to perpetual or periodic war.

Enculturation takes place simply by our being conceived, born, and brought up in a culture and having our mind-brains shaped accordingly. Culture is a mental and emotional force acting on us as a top-down influence along with the bottom-up function of our natural world. We sustain culture simply by living out the neural imprinting culture itself has individually and collectively wrought on us. We cannot be directly aware of this perpetuation because, from conception, our awareness has itself been shaped by culture to an unknowable extent.

Culture constantly acquires variations and colorations that we, as enculturated people, consciously and continually bring about as we grow, adapt to, and mature in our particular cultural environment. Culture is an artificial overlay on our natural system, and we are driven to protect it in an attempt at our own self-preservation, even as we are intuitively driven to escape or change it because it is painfully wounding. We unconsciously live out this contradiction. Our automatic drive to preserve and yet change culture is a paradox that underlies our individual and collective history. We want to change the effects of culture, but because we are identified by it, loss of culture is synonymous in our minds with death. It is our identity, making the cause of culture and its effect simultaneous, each giving rise to the other.

So, paradoxically, our attempt to change culture and its negative effects on us actually preserves culture. Its ceaseless variations are brought about by the constant improvements in culture that enculturation itself impels us to make. The more cultured we are—that is, in this usage, enlightened, educated, refined, and sensitive—the more we aspire to change culture, thereby giving it our strength and sustaining it. In fact, nothing furthers the cultural effect so strongly as this compulsion of ours to change it. In our attempts we project our internal disease onto external causes, thus masking our real dilemma. In our projection, we see that our problems, angst, and frustrations are brought about by phenomena or events of our neighbor or the world out there. We are driven by our defensive survival system itself to bring about the needed change in our neighbor or world as we simultaneously hold to our ideation lest we collapse into chaos—and that ideation is culture.

Further, the lifelong devotion of self to changing culture is held by our cultural-social environment as a noble ideal. Nevertheless, whether vigorously anticultural in our intent and thinking of ourselves as firebrand revolutionists or viewing ourselves as loving, self-sacrificing saints, we serve culture in spite of ourselves.

We cloak our compulsion to change culture under a multitude of disguises, nearly always virtuous: We want to make the world a better place for our children, build a better tomorrow, create the kingdom of God on earth, shape our future through stalwart effort and responsibility, win our wings on the social scene or in the marketplace, be somebody, work for the happy marriage of religion and science, be a real mover and shaker, prove our merit, justify our existence—the list goes on ad infinitum.

Such a nihilistic, pessimistic observation as this seems to leave us with no apparent way to turn or place to go. But note that culture offers endless variations on that very theme of righting the affairs of culture. All these schemes calling for our support are, in effect, an underground run by the establishment, which always leads back to support of the establishment from which it would deliver us.

So, while the content of culture changes continually, through our constant drive to escape it and make life better, culture as a malevolent formative field effect never changes. It is as unseen and unrecognizable as electromagnetic fields once were and simply seems to be the life condition earth affords. Widespread recognition of it as the antihuman, spiritual sickness it is seems to get us nowhere, while we project onto neighbor, fate, nature, God, karma—what have you—its ill effects. Interesting (and irritating to hear), my meditation teacher Baba Muktananda once said we could not recognize defects in another were those defects not first within us.

Consider again that mythical cultural lie concerning hypothetical human nature: Without the constraints of culture, we would be demonic, brutal beasts. This notion has left us with no recourse but to accept culture's principal support and guise, religion, as the only antidote to violence. All the while, religion spawns its manifold form as creeds, bibles, cults, money, politics, power systems, and myriad colorations and cultural flavors and, as an effect, is then sustained through the chronic fear, rage, guilt, and frustration religion produces in us. What results is periodic warfare brought about by the cumulative buildup of social rage and frustration, which is brought about in turn by the limitations and constraints we enforce on each other and our children for their own or the social good.

We pass off this crippling of spirit as the failings of human nature and look on religion as the culturally acceptable solution. Though it is the principle way by which negative and "unnatural" nature is brought about, religion has long been held as the only antidote to the failings and atrocities of human nature. Culture as a force field is not, however, a moral, ethical, or religious force. Its base and its resulting religions are biological, and the only possible cure for its ills rests in the realm of biology. What, then, might break the vicious cycle of culture perpetuated by religion and outrage? The answer is biological, as part 2 here will explore. *Bio* refers to life and *biologic* refers to the knowing, the logic, of life itself, an orderly organization of the forces producing us and available to our common sense because we are of these forces.

Culture can be summed up as an intricate web of admonitions, established commandments, and improvised commands demanding, from our first breath, that we modify our personal behavior to conform to a certain standard. For those who do not conform, culture has an established network of laws of every description concerning guilt, shame, reprisal, punishment, ostracism, alienation, isolation, pain, or death that will be inflicted on us. "Do this or else" is the cultural sword hanging over our minds and thought, and through this is achieved an artificial but semieffective social cohesiveness and a form of behavioral control of a group or groups of people that necessarily leads to periodic episodes of war.

To further explore culture and its myriad forms, we must consider the theory of field effect and understand what is meant by the word *field* as it's used in this sense, for upon this notion turns much of this book's argument. In the next chapter we'll explore the work of Marghanita Laski and her study of the phenomena of discovery and revelation, which involves examples of field effects common to us all, for a grasp of this is critical to understanding the rest of this book.

3

MARGHANITA LASKI AND THE TAUTOLOGY OF FIELD PHENOMENA

Philosopher and writer Marghanita Laski wrote a classic description of the Eureka! discovery or insight occurring continually in science, philosophy, the arts, spiritual paths, and other realms of human experience. I have referred to Laski's outline in nearly every book I've written but am highlighting it this time because of its importance as a shaping force in life as creation—one that shows us how we, as life's expression, create fields by default even as we are created by them.

This creation of fields is a random, amoral (neither good nor bad) process wherein such products of our mind as judgment, criteria, or good-bad evaluation don't apply. Field effect includes not only those fields giving rise to us but also those we in turn give rise to by nature of the whole creative process. Products of this random process include that most powerful and negative force of history: culture.

Examples of Laski's Eureka! experience include William Hamilton's discovery of his famous quaternion theory, a cornerstone of modern mathematics. Hamilton had spent years trying to solve a particular mathematical enigma, with his wife reporting that again and again he grew discouraged and quit his pursuit, vowing he would waste no more of his time on it, only to be caught up in his quest by a new angle he had not thought of before. Finally, after fifteen years of this, he truly quit, lamenting that he had wasted the best years of his life. He asked his wife to take a walk with him to ease his sorrow and, while crossing a little footbridge near Dublin, his mind a blank and finally at rest, the answer arrived. It came in a flash of insight, the entire theorem presented in highly symbolic fashion in a single instant. He later reported that he knew at that moment that the complexity of his answer might require another fifteen years to translate into mathematical terms.

Another example can be found in August Kekule, the Belgian chemist. Having exhausted his patience in the search for a fundamental principle of chemistry he knew had to exist, he engaged in a mindless reverie one day, escaping from his labors and sitting comfortably at his fireplace. Suddenly he perceived directly in front of him the fleeting image of a snake with its tail in its mouth forming a peculiar configuration. He knew instantly this was his answer, though chemistry could do little with snake symbols. After he translated the image into the appropriate language of his profession, the result was the theory of the benzene ring, a cornerstone of modern chemistry. Kekule reportedly said, in an address at one of the occasions honoring him, "Gentlemen, we should do more dreaming."

And then there's the instance of Henri Poincare. He fruitlessly struggled over a particular geometric problem until one day, stepping onto the tram for his ride home, his mind a blank, he perceived the solution in a single flash of insight, opening for us a new field of geometry.

The list goes on; indeed, most fields of human activity involve variations on this general theme. Laski outlined the pattern of this "Eureka!" function:

1. Asking the question. We must first be seized by a passionate question or quest with such intensity that it overrules all other issues or considerations and becomes our focus and principal concern. Such seizure organizes our energy, attention, and intention into a single purpose: to answer that question, fulfill the quest, discover whatever secret lies therein. The nature of the quest is immaterial. Laski's formula is as applicable to an American Indian vision quest as it is to modern science, art, spirituality, or creativity in general.

2. Gathering the materials for the answer. Passionate commitment initiates an equally passionate search for the answer, which is pursued both consciously and deliberately and unconsciously and automatically. Our mind actively screens ordinary daily events, looking for hints, clues, suggestions that might throw light on the subject. We pursue all avenues, research, perform studies, involve all the disciplines that might prove pertinent or related to our quest. Hope springs eternal in this process, regenerated continually with each new possibility and each discovery of a piece of the puzzle. We are driven toward the goal in an exciting, even exhilarating adventure.

3. Hitting a plateau. A third, more sober phase begins as we exhaust our possibilities; though we labor, we can't fit together what we gather. We hit a plateau, a time of stagnation. The luster of our Grail dulls, all our efforts seem in vain, and the goal has eluded us. At this point, we generally quit and try to get our mind off our passion as we pick up ordinary life. It is the fate that befalls most quests. At this bleak point, however, we may suddenly think of some overlooked possibility —and we are off again with renewed energy and zeal. We may hit many such plateaus and fresh starts through such flashes of insight that point beyond themselves.

4. Dark preceding dawn. We come to a dark night. We bottom out, bitter, perhaps, over a wasted life or portion of life. At this point, all hope abandoned, the goal no longer entertained as even a possibility, the answer may (or may not) arrive. At some odd out-of-mind moment, while we think of nothing, the answer may fall upon us unbidden, filling the vacuum of blank thought. In a single flash we "see" something not seen before—and perhaps not having existed before.

This seeing, called Eureka!, arrives full-blown in the mind, not piecemeal, but instead as a single unified whole, perfect and complete in every detail. Nearly always the Eureka! appears in a richly symbolic form, for symbol seems the way to instantly present a great deal of information or some radically new and hitherto unknown possibility, though it generally must be translated into a form that can be shared or comprehended. Like a lightning flash seen only once but that forever changes our mental landscape, this instant vision can't be predicted or determined. When we least expect it, it catches us off-guard and seems a grace or gift.

Apparently the answer can't break into our ordinary awareness when we are preoccupied with our passionate pursuit of trying to find or create it. While we search we are absorbing all energy, clogging the neural networks, enmeshed in the old materials of our quest; the radical newness (or "future flowing into the present," as Robert Sardello terms it) of a Eureka! can't break through. The materials of our search are old even as we uncover them and only replicate the past no matter how we juggle them, which is why we can't just gather up plenty of related material and figure out a synthesis that does the trick. This is part of ordinary problem solving but can't generate the Eureka!

From where does the Eureka! answer come? It is generated through or by the same field of activity generating or generated by the quest itself but may be an essentially nonlocal process—that is, not necessarily located in any specific brain area or even in the brain at all. The answer is almost surely received, however, in the right hemisphere of the brain, the vehicle for perceiving novel information and stimuli for which we have no previously formed structures of knowledge. Our structures of knowing, established concepts, and understandings generally percolate through our left hemisphere, the arena of personal awareness and thinking. But the right hemisphere can bridge the gap to present to the left hemisphere its new potential only at some moment when we are not tying up that left brain in its usual roof brain chattering, that nonstop inner dialogue or flow of idle thoughts. No matter how disciplined, controlled, directed, purposeful, and logical that left-brain thinking is, the mind, in its ordinary sense of self-awareness, must be idle or on hold to clear the decks for a Eureka!

Most thinking that we engage in randomly happens in this way: The left hemisphere's flotsam-and-jetsam feedback from body, world, and memory generally rehashes old events or knowing long after the fact. And as it does this, we spontaneously or unconsciously try to rearrange more to our liking those old structures of memory or knowing or, if on a quest, we try to make all the parts we gather fit together into the answer we seek. This is "the flow of the past into the present," which can only replicate the past. Though perhaps recreated in novel dress, this still obscures the present and the future. There is no Eureka! in that past, while the future flowing into the present and making all things new can open us to the unknown and its potential. Thus, at some odd out-of-mind moment the answer or revelation arcs the gap of our being and arrives full-blown in our awareness.

5. Translating the answer. Next comes a difficult step wherein many a great insight perishes. That unified answer, always appearing as a symbolic whole, generally bears no discernible resemblance to the original question, nor does it relate to any of the materials we may have gathered in our search. This presents the digital-linear left hemisphere, which receives the symbol from the right, with the arduous task of translating it into the specifics of the everyday world. Thus Kekule's snake had to be translated into the language of chemistry, Hamilton's flash of insight into the language of mathematics, and so on. In the realm of spiritual striving, the Eureka! may express as a *metanoia,* or fundamental transformation of the mind, and will have to be lived out for all to see.

Eureka! experience follows the same pattern underlying enlightenment in the life of the isolated monk or nun in a monastery, or the forging of a true mathematical mind or scientist when the student finally sees the world through the particular prism of his chosen discipline. This field effect occurs too in Mozart's creative ability. Such Eureka! moments have changed the face of science, the world, and our worldviews. This experience is an aspect of creation that underlies such diverse pursuits as scientific and philosophical breakthroughs and fire walking in Sri Lanka. Field effect, which gives rise to Eureka!, is itself cosmological, universal, and amoral (not subject to moral judgment). We reap from fields as we sow them, whether the seed we sow is positive or negative. The Eureka!

experience is the exception, unique and unusual, like a tiny pearl found in a field of random clutter.

We initiate the genesis of a Eureka! by our question or questing, and we subsequently bring it about by our passionate pursuit. Yet we can't bring about or think our way into the final answer, that wind of creation that "bloweth where it listeth." All recipients of Eureka! moments refer to their answer or revelation as a gift they had nothing to do with creating, one that simply fell into their mind at some unexpected time. On closer examination, however, we find that while the gift may fall as a gratuitous seed, it falls only into the soil of a well-prepared mind that has resonance with that seed. Otherwise, how could the mind receive, recognize, act on, and nurture that seed to fruition through translation? If Kekule's snake ring appeared to me, I would never in this life have connected it with chemistry. A dynamic interplay or looping effect takes place between the bottom-up activity of our volitional, thinking brain-mind and the top-down action from some nebulous field of potential outside our thought. Mind is the receptor of the process and is critical to it but must be set aside for the arrival of that answer. "We are not the doer" is a refrain I often heard from my meditation teacher Muktananda, but without us, there is no doing.

FIELDS AS FORCES OF CREATION: THE EXPERIENCE OF GORDON GOULD

Physicist Gordon Gould's Eureka! gave the world laser light and won him a Nobel prize, but what is equally significant is how his experience varies from Laski's outline, leading beyond Laski's work itself and beyond most parameters of current thought.

As Gould tells it, he was home for the weekend, relaxing and catching up on odds and ends, when, while thinking of nothing in that typical momentary blank-mind state Laski describes, there fell into Gould's head an incredible image with enormous implications. It lasted only the usual fraction of a second but forever changed the country of his mind. He reports being "electrified, stunned, and awestruck" by its immensity, and he spent the rest of the weekend feverishly scrawling page after page of notes concerning what he had seen in that single flash. On Monday morning, his rough translation complete, he immediately had it notarized to establish his proprietorship of the whole event and subsequent venture.

Like other scientists, Gould puzzled over this bolt out of the blue: Where had it come from? How had it formed? As had others before him, he insisted the vision was a gift freely given, that he had nothing to do with it. More significant for us, he had not sought nor had any expectation or anticipation of that revelation. Further, the gift given was a radical departure from anything he had studied or pursued through thought. Indeed, it fell entirely outside the common domain of scientific thought, for laser light doesn't exist in nature but only through man-made devices generated by Gould's revelation. In fact, Gould's Eureka! involved such a radical departure from nature that it took considerable time and serious effort on his part and the part of several other individuals to translate it into actuality.

On reflection, Gould reasoned that his many years as a physics student and more than twenty years of diligent application of those studies in his various scientific pursuits accounted for the event. He surmised that all these years had, unbeknownst to him, "funneled into his mind all the bricks and mortar" giving rise to the magnificent edifice that had flashed before him in that instant. Although this unasked for answer bore no resemblance to anything in his knowledge or in the various fields of expertise to which his knowledge related, only someone steeped in such knowledge could have both recognized the significance of the vision and translated it into the common language of that field.

The mind here thus proves more problematic than the casual "hopper" referred to by Gould in his account. Missing is the key element in all ordinary Eureka! experiences: No seed was planted, no passionate intent drove Gould in singular pursuit of a goal, no search or gathering of material had taken place. His personal history of diligent work and discipline had given him the well-tilled soil for reception of the seed, but he was dealt an answer that had no question. *Revelation* is then a more accurate term for Gordon Gould's Eureka! It was an effect without cause.

This spontaneous combustion without an igniting spark not only opened a new realm of possibility in optical physics, it opens us to an issue concerning that field process and creativity itself. While the mind draws on and contributes to such fields, in Gould's case an unconscious generative process took place *within that field,* generating as a result a new possibility, a true creation. When translated, his revelation brought about a phenomenon not present in our world. Most Eureka! moments involve the answer to a question concerning some activity or a missing link in a logical structure of thought or even a possibility presented by our imaginative mind. But for Gould, all such preliminaries were blithely passed over and something radically new under the sun was presented or fell into his mind, free of charge.

We can conclude, then, that fields of potential are not only active forces or intelligences within their own domain, they are also creative forces. Gould speaks of the essential ingredients being "funneled into his mind," but we must ask: If that was the whole story, why would his mind have to be suspended or empty for its own answer to present itself? It seems the mind, commonly defined as personal, was not in and of itself the cauldron of creativity forging the revelation. Gould's later insistence that he had nothing to do with the revelation's arrival, in common with all such Eureka! reports, need not be explained away or attributed to false modesty. It is surely true at the individual or personal level of awareness, but the whole is a cosmological-ontological function embracing all minds.

Gould's field fed into a commonly held or universal field of like order, one over and above his or any other individual mind-brain yet obviously the product of his and other minds. The very field of optical physics is itself the result of all physicists involved in optics now or in the past. We speak of going into the field of medicine or the field of architecture or the field of engineering, but *field* here concerns shared social activities and denotes an aggregate of intelligent energy and potential. While field in this sense is a handy term for various actions or states of mind, such a field has no localization. It is, like gravity, a verb, not a noun; a process or procedure, not a product; an aggregate of potential; a hypothetical grouping of related actions. Claiming that such fields can be generative, creating something never before manifested, is bound to be resisted by traditional academics.

Further, fields of knowledge such as mathematics, physics, music, and so on are in a continual flux of arrangement and rearrangement brought about by the constant input of materials from those studying and employing the fields as well as those unconsciously interacting with them. In the case of the field of optical physics, all those students, professionals, amateurs, career scientists, and lonely thinkers mulling over the mystery of light fed into that field the bricks and mortar for any number of great edifices similar to the one Gould discovered.

In this reciprocal action between an individual mind and a hypothetical aggregate of fields wherein a Eureka! can be formed, many individual minds are involved. Gould stood on the shoulders of giants, as did Einstein, Poincare, and the rest, each within their own respective pursuits. No man is an island nor is a field of potential knowledge an isolated phenomenon. There is only reciprocal action between mind or minds and fields.

Because of this, any field contributor might automatically be in line as a possible target for receiving some symbolic answer or Eureka! discovery brought about by that field. The one stipulation seems to be that an individual mind must be idle, vacant, or inactive at the precise instant of the field's creative formation. Out of the ferment of a field of potential interacting with many individual minds, one of those minds may, by chance, be struck by the lightning generated in that particular field—which is not to say that the individual plays no part, for without a resonant mind to receive a field's creative invention of the moment, nothing could happen in either field or mind. Creativity lies in the reciprocal relation between individual and field—which is all an example of Darwin's random mutation and selectivity.

One additional possibility exists: Any one of the contributors to the field might have fed into it a question or seed for a question that acted as that very spark needed for the Eureka! Gould experienced. They may have had any number of issues in mind and may have even sought an answer with passion. The resulting revelation might simply have fallen into whichever mind among the multitude involved happened to be open at the particular instant of creation. In Gould's case, because the Eureka! opened him to an optical phenomenon that had no parallels or similarities in existence, virtually a creation *ex nihilo,* any open optical physicist's mind might have qualified for its reception. What the recipient does with a Eureka! of this nature is another matter, but expectation would have nothing to do with it. Further, he who sowed the seed might well have been left out of the reaping, but injustice doesn't apply. Here in creation's strange loop of cause-effect, our criteria, ethics, judgments, and sense of justice simply aren't relevant. Creation rains equally on just and unjust.

LIGHTNING FORMATION AND EUREKA! EVENTS

The way in which a bolt of actual lightning forms is a powerful analogy of the Eureka! event. Electrical energy generated through cloud activity collects as an attractive aggregate in a particular cloud formation that attracts more and more energy from the surrounding clouds until a saturation point of electrical potential is reached. The laden sky charge then literally seeks out a resonant frequency from a similar saturation point on the ground, which is formed when some small electrical charge in a limited area attracts similar charges scattered throughout the surrounding terrain so that all gather until, in their meeting, they reach maximum saturation point.

This ground charge is always much smaller and weaker than the huge energy generating over the expanse of sky and cloud above, but nevertheless a sky charge with common resonance seeks it out. When the sky charge finds the closest connecting point, the ground charge gathers its full force to make a gesture of attraction toward the cloud charge, using any handy physical object that can conduct it upward—tree, flag pole, elevated structure, or upright human body. This then triggers the far greater sky charge into direct, responsive action. It follows the weaker signal down in a happy rumble on the ground, and in this way, the ground charge has sowed a wind that reaps a whirlwind, so to speak.

Formative fields of potential function roughly similarly, regardless of the nature of resonant energies involved (electromagnetic, chemical, psychological, or spiritual). In Gould's case, when that field of potential hit some kind of saturation point it manifested a new combination of the ingredients within it or perhaps even beyond it, by random chance or highly specific stochastic action. Such combination creation may well be deliberate, purposeful intelligent action over and above any field effect (or our notions of intelligence), though random chance and selectivity are likely part of the process.

The significant factor in Gould's case is that this field of optical physics generated the revelation of a phenomenon not in actual existence anywhere in the universe. Yet that manifestation could have taken place only through or in a properly prepared target: a mind that could attract, receive, and translate that bolt out of the blue. Whether by chance or design, through resonance between Gould's mind and that highly specialized field of mind, creation of something that had never before existed took place, bringing something new under the sun. Creative action of this sort may be a general characteristic of field effect. "Saturated" fields seeking resonant reception may be far more prevalent than the comparatively few cases brought to our awareness.

For instance, I read a report years ago that told of two mathematicians on either side of our globe who came up with the same new theorem at the same time. Fields are nonlocal, but they may localize in some brain or brains at each manifestation. Where those brains are located doesn't enter the equation. Further, such action clearly displays an intelligence at work. To deny that intelligence plays a part in the Eureka! process giving rise to, for instance, mathematical breakthroughs is to deny that mathematics involves intelligence.

The analogy of the lightning bolt carries additional weight if we can recognize that all field manifestation is of the same natural order. Biologist Gregory Bateson claimed that mind and nature are a unity, and whether a field that forms from our action is of mind matters such as mathematics or music or from nature matters such as electromagnetic energy or lightning bolts, both categories encompass natural phenomena and function in the same way.

This chapter has served to point up what fields of mind may do on their own. In looking at the Hubble space telescope's vast stellar clouds in a constellation of stars, we are looking at potential embryos for future suns, planets, and eventual life. Random thoughts in our heads occasionally produce a new idea that we can execute and make real. Should it take a million random thoughts to hit on one that can ignite a truly new evolutionary and transcendent event, nothing has been wasted and all gained.

Consider that the nucleus for a field may be hatched in some lone thinker's mind, catching up other minds in the possibility built up by many participants over stretches of time. Aggregates of knowledge and possibility may go through cycles of growth and maturation, creating continual combinations of possibility, many of which go nowhere in our pragmatic, utilitarian sense. Some ideas may simply dissipate from neglect, attracting no lightning from other clouds, while others may be carried through by the appropriate mind vehicles. Thus some brilliant and informed individual may, on a clear day without a cloud in the sky, be struck by a Eureka! lightning bolt that sparks real thunder, changing the country of his or her mind and perhaps those of many other individuals—provided, of course, that the receiver serves the revelation in a translation that completes the looped action.

Laski points out that sometimes the recipient of a field manifestation is not up to its translation or that there exists no suitable ground for such a translation. The event then aborts, unfulfilled at that time. It is interesting to consider that the Eureka! event may be happening throughout the universe on many different levels, a constant creativity randomly producing and selectively expressing as conditions allow. Darwin's random mutation and selective procedure may have always been taking place throughout the cosmos, as it takes place continually in the constant ferment of thought boiling up in our head.

While Kekule suggested, "Gentlemen, we should do more dreaming," William Blake says that great people imagine the highest and most divine, for imaginations are divine, and anything capable of being believed is an image of truth.

4

MIND AND FIELDS OF MIND

Princeton University psychologist Julian Jaynes wrote of a particular pianist's ability to produce cascades of notes at high speed while apparently thinking only of the nuance, shape, shading, or expression of that which he wanted to hear. Jaynes, a pianist himself, explained that were the pianist suddenly to become consciously aware of individual notes or feel that he must consciously execute those notes and move those muscles by his own volition, he would be unable to play. Over a half century ago, Ernest Edwards, a brilliant young pianist I knew (on his way to what appeared a great career when he was stopped by arthritis) said he could play no faster than he could hear (though it seemed he could hear at a breathtaking speed).

I have heard pianists play Chopin etudes written primarily in sixteenth notes at metronome settings of 168 to a quarter note. Because the left and right hands play different lines of notes simultaneously, this tallies up to roughly 1,304 individual finger strokes per minute—a bit over twenty per second. Julian Jaynes would ask: How are those notes being executed? Through the mind? Through consciousness in the sense of self-consciousness? Hardly, it seems. If any of these mental factors enter the scene, playing stops. Something different is going on, but what?

How is it that we must hear in order to execute that which is heard? What is it that comes first? Could it all be related to the cause-effect mirroring of the strange loops we have been examining? We assume that when we move our arm, some part of our neural system informs another system through intricate neural feedback loops, an assumption that breaks down under close scrutiny, as Robert Sardello asserts. Jaynes points out that consciousness, at least self-consciousness, is not necessary for most of our actions. We seem to furnish intent often enough, but what carries out that intent is less clear.

Back in the late 1970s, my meditation teacher Muktananda reiterated many times: "You are not the doer." In response to our question "Then who is?" he replied patiently: "The same Shakti who creates your world." This Shakti, according to Eastern theory, is a field of creative energy in which we are nested. Brain-mind and actions are inextricably intertwined, and while we have some understanding of brain, we are less clear about mind. Who is writing this line or eventually reading it except a mind with a brain or vice versa?

Gerald Eddleman's book *Neural Darwinism* explores how neural fields of the brain form. These fields are aggregates of anywhere from a mere hundred thousand to a million or more interconnecting neural cells, each aggregate forming to translate a specific frequency of energy. Countless numbers of these fields form in infancy and childhood, but they continue to form throughout our lifetime. All the myriad translations from all the translating fields of neurons interconnect somehow to form the whole perceptions we experience as our world. Yet no one can find a place in the brain where such a connection or actual world experience forms.

NEURAL ENERGY, MIND FIELDS, AND THE TOP-DOWN FIELD EFFECT

Neural fields translate from energy fields, but where those energy fields form into aggregates and how they interact with neurons is all very fuzzy. Field theory concerns aggregates of resonant energy, groupings of similar frequencies or vibrations or waves of energy that are compatible—that is, that are resonant and can cohere, mesh, or match through peak and trough, at least periodically (see fig. 4.1). Such fields of energy underlie all structures or events in our experience, whether the structure is of thought or matter, and suggest a top-down influence in the reality we experience, whereas conventional science admits only to a bottom-up explanation.

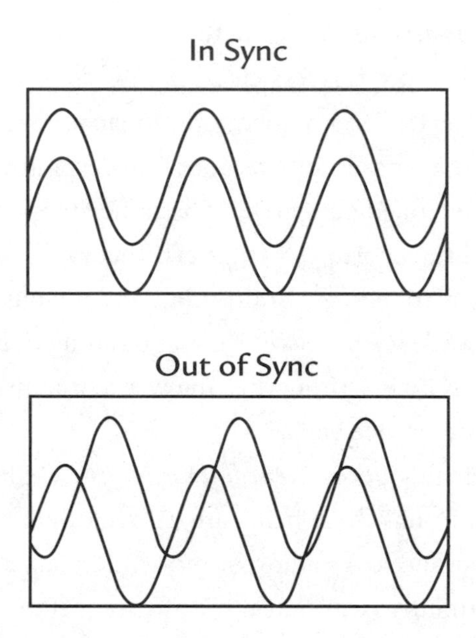

Figure 4.1. Coherent waves can match peak and trough and reinforce each other. Incoherent waves clash in a dissipating static.

A bottom-up explanation refers to the dogma that all phenomena arise from an atomic-molecular substructure of some irreducibly tangible "stuff" that is in some way quantifiable in a laboratory. This notion was reinforced by an equally hypothetical big bang theory, which scientists claim gave rise to the universe and to which all bottom-up notions can be traced. In many ways, this belief structure is similar to that espoused by the Bible-quoting fundamentalist, and if you don't accept the basic premise on which the belief is based, you are simply deemed a heretic whose observations are dismissible.

A top-down influence arises from a field effect. Introduced back in the 1920s in relation to memory, learning, and heredity, field theory was picked up decades later by Rupert Sheldrake and David Bohm. Sheldrake refers to morpho-genetic fields underlying all our reality. (*Morpho* refers to "form," while *genetic* refers to the genesis or origin of form.)

Harvard's Howard Gardner developed a theory of multiple intelligences that lends itself to or indirectly supports field theory. We are born with or develop a general capacity for learning various disciplines, but Gardner proposed that each major discipline we learn, such as music, mathematics, language, or spatial concepts, is essentially a discrete, separate potential capacity within itself, a universal template out of which arises our individual employments of it. We must, of course, selectively focus on and develop such fields or, in effect, lose them.

In the last three decades or so, the evidence for field effect has accumulated, supporting the notion of a top-down influence that falls outside the boundaries of conventional science, which is obligingly, if slowly, changing accordingly. David Bohm proposed that consciousness itself was a field of energy underlying all possible arrangements making up a universe, and considerably earlier the Swiss philosopher Jean Gebser presented a major opus based on an "ever-present Origin" from which all events spring, including any hypothetical big bang, whether in the past, present, or future.

In the early 1970s neuropsychologist Karl Pribram proposed that the brain-body puts together our picture and experience of reality by translating frequencies that, instead of being in time-space, are generated from a spectrum of energy that apparently gives rise to such frequencies. Such a spectrum has no locus; it is simply a potential. This is a top-down proposal that has gained support since, but that was not smiled upon by academic bottom-up believers on its introduction.

Fields of potential express through neural fields and are perceived by our mind as our reality. Mind arises parallel to and from the actions of such neural energy fields and continually reinforces such fields by its interaction with them. All this reciprocal action is brought about by a consciousness that, according to Bohm and many physicists following him, is the field of all fields. Yet any field, including even a hypothetical universal one, is brought about and sustained equally by actual minds and brains, yours and mine, one at a time or all together.

MUSIC AND THE MYSTERY OF MIND

When given a commission for a piece of music, whether quartet, symphony, or concerto, in his mature stage of musical development Mozart would reportedly simply sit on the request, holding it in the back of his mind but not actively thinking about it. Often as not, at some point the musical piece would visibly and aurally manifest to the composer in its completeness in a single moment (akin to the Eureka! experience).

Mozart spoke of this visual and auditory event as a "round volume of sound" in which every note, nuance, and shading was presented in the context of perfect formal structure. He then had the formidable task, however, of translating the whole into its myriad inner parts, all of which he did in his head. After this, he had the equally laborious task of translating those thousands of sounds into individual inkblots on paper that other musicians could read—by which method he would re-create what he had heard in that out-of-mind moment. Often Mozart would have his wife read to him or he would sip hot chocolate and eat cookies to occupy his mind as he wrote out those thousands of notes. So which part of him was doing what and when?

A pianist friend once told me about a particular concert in which he was scheduled to play his favorite Mozart sonata, one he considered the most perfect and beautiful musical work ever created. Right before beginning, he closed his eyes and leaned back to immerse himself in the spirit of the piece and his feeling for and love of it. It was then that he experienced in a single second the entire work as, in his own words, a "round volume of sound" in which every line, bar, and note was complete in its most pure and pristine form. It was, he recounted, a mystical, spiritual, transcendent revelation that resulted in the greatest playing he had ever done.

In his book *Godel, Escher, and Bach,* Douglas Hoffstadter recounts the improvisatory capacities of J. S. Bach at the keyboard—his ability to spontaneously create music on demand. Legendary is Bach's improvisation of a six-voiced fugue on a theme furnished by his benefactor, Frederick the Great, king of Prussia. The structural technicalities of a fugue in music composition are tricky to begin with, and such an accomplishment as Bach achieved on that occasion lies beyond the ordinary reaches of mind. According to Laski's analysis, Bach's feat may have arisen from a field beyond mind itself (if mind refers to only an emergent property of a single brain in a single skull).

In the middle of an interview with Artur Rodzinsky, a famous orchestra conductor of the mid-twentieth century, a reporter asked if Rodzinsky knew the length of a particular work the orchestra would perform that night. "Well," the great conductor said, pulling out a stopwatch, "I can tell you shortly." He clicked the device, lay it on the table, and continued conversing about symphonies, composers, and the like. In the middle of a sentence he broke off, stopped the watch, and reported the precise length, to the second, of the work in question, to the reporter's mystification. Had Artur somehow played the piece in his head without pause during his verbal discussion?

Interestingly, recent brain scans of active music listening or performing reveal that the same areas of the brain light up when we hear a piece of music in our heads and when we hear the piece actually played. These areas, involving large portions of the brain, are not the same as those used in speech. So if, as it seems, Rodzinsky did listen in his head while talking, which activity involved his mind? As my editor pointed out to me, given that the portions of brain involved in listening to music and speech are not the same, then an extraordinary person could do both at once. And indeed this is a grand example of the mystery of mind we can't explain away.

A related though nonmusical example can be cited from World War II in the signal corps, which trained young men to send and receive wireless messages by a code similar to the Morse dot-dash system of the early telegraph. One particular young corporal was the fastest telegrapher in the Sixth Marine Division on Okinawa and could receive messages in code and translate them in his head faster than he could physically write down those translations. Often he would continue writing for some time after the machine had shut off: All incoming messages had been received, translated, and carried in the boy's head but not yet recorded on paper. The brain-mind was faster than the brain-hand, the receptor was faster than the executor. To add to the feat, as my brother in the same Marine squad reported, the whole time he was working out code, the boy was likely to be carrying on conversations with the other troops sitting nearby or dodging bullets in his foxhole.

Arthur Rubenstein, one of the great pianists of the twentieth century (and a great human soul) provides yet one more example of the mysteries of the mind: A composer friend gave him a new manuscript to look over on the long flight to his next concert. Rubenstein indeed looked over the new piece, was quite impressed by it, studied it throughout the flight, and played it that night as an encore—and by memory.

A CASE OF MUSICAL GENIUS

Around 1948 in Atlanta, Georgia, friends brought over to my place a young neighbor of theirs, a boy of fourteen, who had asked for their help in finding a piano teacher. It seems the boy had been taking lessons for a month or so from the piano player at the local Four-Square Gospel Church. Old-fashioned shape-note hymnals were all that was available to him as teaching texts (and also comprised the extent of the teacher's musical knowledge). These hymnals had been the boy's only exposure to printed music, and after a few sessions of such limited fare, he looked for wider horizons.

I was one of those all-thumbs people who longed to play the piano but whose right hand never knew what the left was doing and never would know, yet I had a formidable stack of music I loved to stumble through in my slow hunt-and-peck fashion. Somehow, just buying volumes of Beethoven sonatas or Brahms intermezzi was satisfying to my soul, though my painfully slow working out of which finger might be called for each note and writing these numbers above the notes themselves to aid my efforts meant a long lapse before I could get a reasonable, albeit slow-motion, idea of what the composer might have intended.

Yet I loved sound for the sake of sound and harmony for itself, which somewhat made up for the missing rhythm, tempo, and flow of phrases. I could also often manage to play fragments of recognizable themes if I isolated each hand and played their parts separately. Interestingly, while I was so slow at playing written music, I could improvise fairly well, particularly in the dark, and would do so for hours, thundering out emotional storms brewing within me, crowding to be expressed, and furious rhythms that shifted and charged about. I couldn't do this if I thought anyone was listening, however, and I still couldn't sight-read a simple hymn—and I can't to this day. Some part of my apparatus was simply missing, brought about, I later concluded, by my brain's left-hemisphere dominance.

At any rate, my friends brought over their young charge and asked me to see what I thought of him, for apparently he could sight-read any piece of music. I started him off with some simple introductory pieces that he tossed off at a glance. Progressing to more and more challenging works, he seemed to grasp huge handfuls of notes as he effortlessly thundered them out on the keys. I finally put before him Chopin's "Winter Winds Etude," a great tangle of accidentals, sharps, and flats thrown in en masse by the composer in a chromatic torrent designed to be played like the wind. It was a formidable challenge to read and play at any speed. The boy caught the edge-of-madness spirit of this wondrous work and, without hesitation, made live, dramatic music of it, leaving me dumbfounded and perplexed. I then brought out the Busoni piano arrangement of the great violin Chaconne in D by J. S. Bach, a truly challenging epic, and he tossed it off, sensing (it seemed to me) the profundity of the work.

All this was beyond my comprehension, and I called a top-flight pianist and teacher friend and explained the situation. He came over right away, puzzled at my exuberance, and began running the boy through the literature, ending the afternoon walking up and down the room and weeping while saying that he had never before encountered such a gift.

William Blake said, "Mechanical excellence is the vehicle for genius," but where in this boy were all the years of preparation supposedly required to produce that mechanical excellence displayed in his genius? I have concluded this: *It* (in the original Zen sense of the word) breathed that young man, with *It* in this instance being the field of music. Apparently this field can function within the neural field of anyone able to respond to it. I found nothing in this boy's background that suggested a possible reason for this gift. It is interesting to note that the clues concerning mind in this example are related to the clues existing in those highly gifted autistic individuals who have what is known as savant syndrome.

I lost all contact with the boy after I left Atlanta that year, and several years elapsed before I heard the final sad chapter of his history. Apparently his wealthy parents had split in a nasty divorce and neither wanted the boy. The father "lost" the court custody battle and had to take him into his own household, which now also held a new and jealous young wife. The father built the boy a wing on the house to which the child was confined, apparently with his piano. A year or so later, the story goes, the young man was dispatched to the notorious Georgia State Mental Hospital at Milledgeville, source and backdrop for the terrifying book and film *The Snake Pit.*

Hearing this, I realized anew the lie in that old saw "genius will out." More accurate might be the phrase "genius can easily be snuffed out." Like great intellect, genius is a fragile phenomenon and can be lost in a number of ways. In fact, if you dig deep enough, the greatest hurdle that genius must overcome or fall on is culture itself.

MUSICAL INTELLIGENCE

What all of these accounts suggest is that there exists a musical intelligence that seems to function outside the ordinary boundaries of an individual mind. Similarly, as displayed in a superior pianist, there is an intelligence of the body that can manifest beyond the confines of conventional thought and be incorporated into a higher function of field effect. Apparently, without the mind's help, this intelligence moves muscles through neural networks not available to our analysis. So the mind must in some sense be suspended in order for the field to fully express, precisely as in Laski's Eureka! effect. This points up the paucity of current neuroscience explanations for such phenomena and brings us back to Julian Jayne's questions about who is doing what.

Through the ages, music has evolved into a formidable field of potential that can rally behind any student's efforts or, in some cases, simply take over the show or that can spontaneously bring into being a new creation that has simply fallen into a composer's mind much like Gordon Gould's laser fell into his.

Mozart and Bach had achieved a mechanical excellence that few musicians ever had, but the genius that flowed through them may have been beyond even the excellence of those minds that conveyed it. Meanwhile, the young man from Atlanta throws a monkey wrench into many of our commonsense observations. Because of abilities like his, we must go back to the drawing board for our notion of how to attain that mechanical excellence necessary for genius to flow. At times, such mechanics may arise from the same field to which such abilities apparently open us, showing mind and its fields of potential to be reciprocal interactions, which reminds us of Rudolf Steiner's reference to "knowledge of the higher worlds."

IDIOT SAVANTS

In my writings, I have often made reference to the savant phenomenon that seems to embody so much of the mind's mystery and knowledge of higher worlds. Recently I was sent a dispatch from the British journal the *Guardian* concerning a twenty-six-year-old savant, Daniel Tammet, who "can't drive a car, wire a plug, [or] tell his right from left," yet working with investigators, as he is always willing to do, he can perform calculations quicker than a calculator. When asked to multiply 377 by 795 Tammet knows the answer in a blink because, as he himself explains, he doesn't calculate; he sees numbers as shapes, colors, and contours that form instantly in his mind, in what he calls "mental imagery, maths without having to think." Along with other implications, this brings to mind a recent theory in neurology that suggests the brain communicates within its modules and lobes through images. Under laboratory testing with an adjudicator present, Tammet recalled pi to 22,514 decimal places—which took some five hours to do but that he gladly performed to show, as he explained, that though he was handicapped, he wasn't stupid. On another occasion he recited those 22,514 numbers backward.

Tammet also picks up other languages ad lib; speaks French, German, Spanish, Lithuanian, Icelandic, and Esperanto; and has invented an extremely complex language of his own that is syntactically logical, coherent, and consistent. Professor Simon Baron-Cohen, director of the Autism Research Center at Cambridge University, points out that he has studied other savants who could speak several languages, but none who could discuss their abilities, much less invent a complex language of their own. Trying to discern which brain modules are involved, researchers at Cambridge University have been performing brain scans of Tammet. Apparently, in savants with his abilities the right hemisphere plays a greater role than it does in ordinary people. Tellingly for our purposes here, Baron-Cohen wonders why such abilities couldn't be available to everyone. We might all then, it seems to me, attain higher worlds of knowledge.

Revealing still more about the phenomenon of savant sydrome, an article in the December 2005 edition of *Scientific American* explores the abilities of Kim Peak. An intriguing sidebar to this article is that academic science can indeed slowly shift its attitudes and become more open to phenomena that call to question previous academic assumptions: The phenomenon of savant autism was explained away or generally dismissed twenty years ago. Darold Treffert—who, in 1988, published one of the earliest serious studies on the subject of savant autism in his book *Extraordinary People*—presents in this article an open-minded scientific approach to the savant enigma.

Kim Peak's specialty is memory in general, rather than just a specific ability such as knowledge of calendars and dates, or an exceptional ability with numbers, as is usual with savants (and which Kim also displays; he knows all the area codes and zip codes in the United States and the television stations in all these locales, and he can cite the traveling distance between any two cities). Kim began memorizing books at eighteen months of age as they were read to him. To date he knows nine thousand books by heart and has available recall on all of them. He reads a page in eight to ten seconds, recording the page in his mental hard drive as he goes. He has a wide variety of interests and an encyclopedic knowledge of each, including history, sports, movies, geography, space programs, the Bible, church history, literature, Shakespeare, and classical music. He can identify hundreds of classical music compositions, tell when and where each was composed and first performed, give the name of the composer and his or her biographical details, and discuss the formal and tonal components of the music. Further, he can identify the composer of a piece not heard before by assessing the musical style involved and deducing who the composer might be.

In his fifties Peak picked up piano playing, an adjunct to his remarkable memory of music in general. In one meeting he presented the opening of Frederick Smetana's tone poem "The Moldau" by reducing the flute and clarinet parts to an arpeggio figure in his left hand and introducing the oboe and bassoon parts of the primary theme, playing them in thirds with his right hand while his left continued the arpeggio figures as in the score—and all of this was done by memory. He also has tremendous associative capacities ordinarily missing in savants.

Kim walks with a peculiar sidelong gait, can't button his clothes, and can't attend to simple tasks. His cerebellum, which controls movement and speech, is quite deformed and he is missing several brain parts considered critical, most notably the corpus callosum, which connects the right and left hemispheres. Darold Treffert speculates that Kim's right hemisphere has had to assume many tasks ordinarily handled by the left. Interestingly, there are some similarities between his mental operations and those of Mozart (who also had a very large head and less practical intelligence). Treffert and other psychologists studying the savant phenomenon point out that we may all have capacities of this general order, but most of these are ruled out by overdominance of the left hemisphere, a notion I proposed a number of years ago from my own personal experience and observations of my children in their development. We might wonder if this hemispheric overdominance is the result of the behavior modifications enforced on us in early childhood to assure our cultural conformity, a thought that will be explored more fully in part 2 of this book.

Rupert Sheldrake claims we create fields of potential when some action is repeated by enough people over a wide enough area enough times. Any new invention, such as computers, once embraced by group activity, may bring about the equivalent of its own morphic field. I faced a severe learning curve when I switched from my ancient portable typewriter to my lap-top computer in 1985. Today, two twelve- and thirteen-year-old neighbors of mine build their own computers from parts gleaned from junked models found in dumpsters and have kept my own lap-top as functional as that of any professional. (As an aside, they were both homeschooled in a casual haphazard fashion and are now in college.) They have apparently acquired their computer knowledge by osmosis, picking it up from our technological ambient itself, perhaps an action involving mirror neurons. In this way, a field effect or possibility builds up its potential, which then can mirror, lend to, and enhance any action of ours that is resonant with it. To him who has, more is given: in the reciprocal mirroring between mind and field, each can amplify the other.

Fields of intelligence may lie outside time and space even as they factor into the nature of time, space, and the intelligence of life itself. So, while field intelligence involves brain-mind, I doubt we can equate field intelligence with mind or mind with that field or even with brain. Yet all three—mind, brain, and field—are interdependent, each giving rise to the other in a looping effect that underlies experience. Any field effect, however it is created or formed, is simultaneously independent of any particular human system even as it is equally dependent on some actual human system for activation, sustenance, and expression.

We can see how this notion of field effect neatly dovetails with Marghanita Laski's Eureka! outline. Apparently we tap into such a field effect when we set about to develop a particular talent or pursue a new interest, whether music, mathematics, art, water dowsing, or juggling. What's more, every activity in such an area, no matter how superficial or in-depth it may be, strengthens and perpetuates the corresponding field.

Here, in all these reciprocating energies and systems, lies the power and stability of culture as a negative field effect, though the same reciprocal function could have just as easily established a positive, peaceful, and evolutionary mind-set. That it may still establish this positive effect is not as far-fetched as we might believe. Hope can be a powerful weapon for good just as doubt works for evil, setting up a light-dark polarity and struggle that Rudolf Steiner explores in *Approaching the Mystery of Golgotha*.

5

MIND AND INTUITIVE PERCEPTION

Man's perceptions are not bounded by organs of perception; he perceives more than sense (tho' ever so acute) can discover.

WILLIAM BLAKE, *There Is No Natural Religion*

In 2004, discoveries at the Institute of HeartMath, a heart-brain research center in the Santa Cruz mountains of California, revealed that we have within us a clear intuitive sense. Our heart, researchers there found, senses the nature of particular events ahead of time, before these events actually occur. The heart clearly signals the brain if an event about to manifest is negative in nature or is one to which the individual will react negatively. This newest discovery falls outside our current parameters of accepted possibility and even those determined by the Institute of HeartMath up to now. Besides revealing an intuitive intelligence within us, it implies a distinction between brain and mind.

At the Institute of HeartMath, a person wired for brainwaves (through an electroencephalogram, or EEG) and the heart's electromagnetic (EM) spectrum (as recorded by an electrocardiogram, or ECG) sits in front of a viewing screen. When the viewer's heart-brain biofeedback devices show these systems to be settled and calm and the viewer feels ready, he or she pushes a button. Ten seconds later a random selector clicks in, instantly selecting and displaying on the screen a picture drawn from several dozen pictures in a computer's selection pool. As with all random devices, no one can predict which picture will be selected. Some 20 percent of the pictures in the available menu are repulsive, chosen for the menu because of their negative nature. The rest of the pictures are benign, chosen to elicit no negative emotional reaction.

Four to seven seconds before a negative picture is randomly chosen and presented, the heart-brain feedback systems register a distinctive pattern of response. The heart responses made to negative scenes at this early stage, before the pictures actually appear, are markedly different from those heart patterns occurring before benign pictures are shown. Thus the study has determined that the heart-brain clearly indicates knowledge of a negative event well before the event physically materializes. In this predictive function, the heart response precedes the brain shift (located in the frontal lobes of the brain) by a fraction of a second, and after a split second, the frontal brain shifts simultaneously with the heart, indicating that heart and brain interact.

The HeartMath experiments were originally set up to further investigate various well-known aspects of sensory perception. The predictive capacities displayed by the heart were an unexpected peripheral effect that opened up a new branch of research. The Institute of HeartMath ran 2,400 trials before they released their first paper on this intuitive capacity and they have since continued with their research.

What is most significant here is that the wired viewer is unaware of his or her heart's four-to-seven second anticipatory activity or predictive function, and is only aware of the actual appearance of the image on the screen. Therefore, it seems, there is a heart-brain awareness in us that oddly precedes our mind awareness. There are countless anecdotal reports of this heart-brain awareness, which, of course, carry no weight in the world of academic science, but can tell us a great deal. In exploring the question "What is mind?" this intuitive capacity is quite pertinent to our understanding. What it shows is that in the case of intuitive awareness, the mind is the last to know.

ORIGINAL WISDOM

The fact that our human brain was added to the kind of brain structures found in all mammals and that our mammalian brain overlays a reptilian brain gives some validity to the process of studying animals to get some idea of what makes humans tick. To a point animals can tell us what bearing these older brain structures have in our own system. Intuition, for instance, which might be thought of as a faculty of very high intelligence, is actually found throughout the animal kingdom.

More than fifty years ago I came across an account from a naturalist concerning intuition in wild animals. His most vivid story was of his daily observations on the progress of a mother fox and her cubs, secure in a deep den several feet up the bank of a rushing mountain stream. The kits were finally beginning to appear for short stretches when mama fox would bring mice and moles to them. One balmy, clear day, when a high breeze was blowing, that mother fox did something the naturalist had never seen before and that briefly baffled him: The vixen suddenly emerged from her den, scrambled some fifteen feet up the bank, well above her home, and, with dirt flying, began furiously to dig another den. In a short time she had disappeared into the new hole, still digging away, and finally, reappearing, she ran down to the original den and laboriously carried each little kit to their new home, where she safely deposited them.

Within minutes after completing this unusual feat, a flash flood tore down the steep mountain valley, creating a wall of water that carried masses of debris. The former den flooded and remained submerged for quite some time until the jammed water subsided. It seems a massive cloudburst had loosed several miles upstream, unbeknownst to the naturalist on that balmy day but not to the vixen.

Following the great Asian tsunami of 2004, caretakers of several of the wild animal reservations in the island countries of Sri Lanka,

Malaysia, and the rest of the area reported that some ten minutes before the great wave struck, the animals in their reserve all stampeded away from the shore and to the highest ground. A similar account concerned a group of elephants kept by a luxury tourist hotel on a Malaysian beach and trained to carry visitors on a sight-seeing tour through the jungle and along the shore. The beasts were secured near the hotel, each with one leg tethered by a chain linked to a post driven deep in the ground, not so much to keep them from wandering but to reassure the tourists who clustered around to feed them and gawk. Several minutes before the tsunami struck, the elephant that was then on his trek and near the beach suddenly turned and rushed back to the tethered group, trumpeting loudly, much to his driver's and passenger's alarm. The maverick then began helping each elephant pull its tether out of the ground, after which they all rushed noisily to the highest point of ground in the area. Many from the native population rushed after them in alarm just as the great wave struck.

In recent years, Rupert Sheldrake has published documented, quantifiable, and repeatable tests and experiments concerning intuition in animals. He recounts many displays of striking intuitive intelligence in a number of animal species. Two of his books, *Dogs That Know When Their Masters Are Coming Home* and *The Sense of Being Stared At,* are filled with anecdotal accounts and a wealth of specific laboratory accounts of animals' intuitive capacity, which he attributes to the morpho-genetic field theory he has explored for years. Taking his studies one step further, the field of neurocardiology and the Institute of HeartMath offer a wealth of concrete, tangible biological patterns that explain both morphic fields and the way in which they are shaped by and in turn shape our reality experience.

Anthropological studies of such people as the Kalahari !Kung, Malaysian and Australian Aborigines, as well as some Eskimo show strong intuitive capacities traceable even today among the scattered remnants of such societies. The famous explorer, anthropologist, and author Laurens van der Post wrote a moving account of a group of !Kung dying of starvation and thirst in the Kalahari desert. Van der Post's guide and driver was a !Kung who had left his ancient ways to team up with van der Post, but his intuitive heart intelligence, an indelible characteristic of the !Kung, remained intact.

One day, having found a clump of desert trees and made camp for the hottest part of the day, the !Kung driver informed Laurens that he heard a group of wandering tribesmen calling for help. "They have had no food or water for days," he told Laurens, "and we must prepare for them." Laurens had heard nothing, but his driver's insistence was unrelenting, so Laurens climbed a small hill nearby and looked in all directions. No one was in sight. "They are getting closer," his driver exclaimed, "there are five of them. Can't you hear them?" Laurens could not, but before long they came into view, five scarecrow survivors, barely alive. "How did you know so far in advance?" Laurens asked. "I heard them," said the driver. "But why couldn't I?" asked Laurens. "Because I heard them from my heart," was the simple reply. And the !Kung are not the only such group to operate with such intuition. Australian Aborigines as well as the Malaysian aboriginals described by the Dutch psychologist Robert Wolff in his book *Original Wisdom* all show striking capacities of this nature, adding up to a voluminous history that is hard to dismiss as simply anecdotal.

THE MIND IS THE LAST TO KNOW

In October of 2001, I and about two hundred people from around the world attended an educational conference held in Hawaii. Among the attendees was a small group of teenagers, each of which was selected to attend based on an essay he or she had written. At this conference a gentleman from England was present with the latest HeartMath apparatus for recording brain-heart feedback (an EEG and ECG). A computer projected the results of this biofeedback onto a large screen so that the audience could follow the proceedings as they unfolded.

Selected as the subject of a demonstration of those functions studied at the Institute of HeartMath, a teenager was wired up for brain-wave and heart-spectrum analyses and sat on the stage facing us with the computer projection screen behind him. The Institute of HeartMath representative talked and joked with the young man, putting him so at ease that his heart and brain went into entrainment, which occurs when the frequencies of brain and heart become synchronous, matching peak and trough (see fig. 5.1). (For years the Monroe Institute has brought about such entrainment or synchrony between the hemispheres of the brain, a powerful state that opens the brain to fascinating phenomena; see fig. 5.2.) The HeartMath representative then turned the young man around so that he could view his own entrainment, explaining to the student that such synchrony of heart and brain is not easy to achieve, asserting that the young man was both bright and well-balanced.

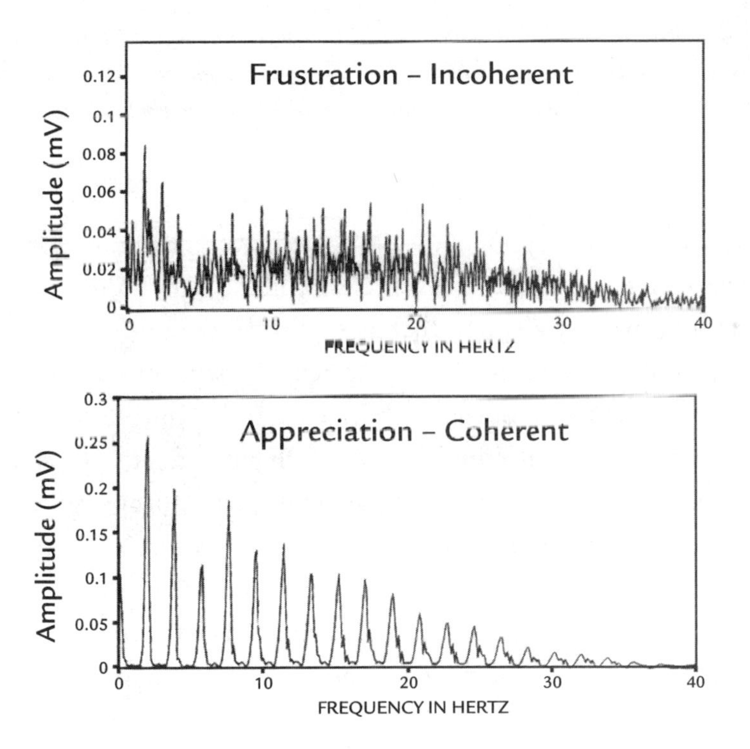

Figure 5.1. Note the difference between the heart's frequency spectrum in a negative state prompted by fear or frustration and in a positive state prompted by appreciation, love, or empathy. Negative frequencies are incoherent—that is, they can't match, or sync, and will clash—whereas positive frequencies are coherent, automatically matching, and in sync so that they reinforce each other.

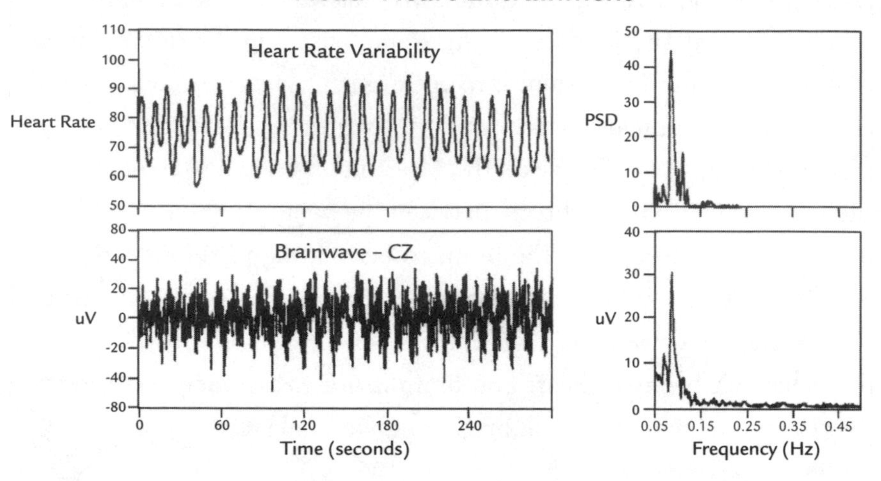

Figure 5.2. Head-heart entrainment can bring a single unified wave form in which the two reinforce and strengthen each other and can then entrain the body as well. With this body entrainment, what we think and feel and how we act become a unit, producing a state of wholeness.

At this point the representative observed that in his state of balance and attunement, the young man was clearly in the right frame of mind to respond to some of the new verbal math quizzes currently used in England. Instantly, as displayed on the screen, the student's synchronous, coherent heart-brain frequencies dropped into the incoherent chaotic mode (see fig. 5.1), reflecting a series of flight-or-fight defensive patterns. We, the audience, burst into laughter at the abrupt shift, which brought only a puzzled frown to the young man's face. "What's up?" he asked, bewildered at our laughter, whereupon the instructor again turned him to face the screen to see the display of chaos where there had been a rare order. That the young man was unaware that a major shift in neural patterns in his own heart-brain had taken place is of profound significance. It was the same unawareness exhibited by the people in the laboratory tests on intuition at the Institute of HeartMath.

The big questions: Why is the mind the last to know? Why are we not aware of what is going on in our own brain-heart dialogue and why are we not part of the discussion? There is apparently no direct equivalence between heart and mind as there is between heart and brain. This, in turn, indicates there is no direct equivalence between brain and mind, though we have automatically assumed there to be.

A piece of the puzzle might lie in those animal reports. They suggest that because they seemingly do not have a mind as we have, animals are aware *as* their brain rather than as a mind *with* a brain. Mind with a brain can, in effect, cut itself off from or be disconnected from the heart-brain dialogue.

But there is more. Recent studies in the brain's blood flow show that when that young man's frequencies registered his alarm at having to tackle verbal math quizzes, the bulk of his brain's energy moved from the forebrain to hindbrain structures. Our forebrain is involved with advanced evolutionary abilities, intellect, imagination, and creativity (necessary to such actions as math problem-solving), while the ancient hindbrain system is associated with reactive and instinctual survival-defense patterns (see fig. 5.3). Thus at the very time the young man

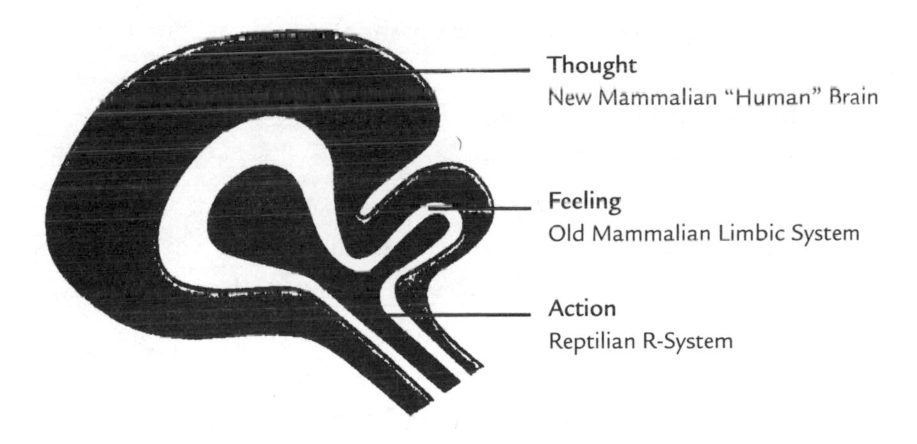

Figure 5.3. The new mammalian brain, associated with thinking and language, and the old mammalian brain, associated with emotions and memory, make up the forebrain, while the old reptilian brain, including the spinal column and peripheral nervous system, is the instinctual, defensive hindbrain, which enlarges in emotionally conflicted pregnancies when a troubled environment is anticipated. Note that in this early MacLean drawing the prefrontals are not distinguished as differing from the new mammalian or thinking brain.

needed his higher intelligences, unbeknownst to him he had far less access to them. Even if mind were considered just a brain function, no matter how specialized the brain parts involved, this couldn't account for the extensive delay between one neural system and another, much less the complete absence of such communication. Brain communication between its modules is nearly instantaneous, hardly subject to four to seven second delays.

THE TWO-FOLD ASPECT OF MIND AND THE MIND AT LARGE

Another wrinkle in our notions of mind came back in the early 1980s from English neuroscientist John Lorber. Using the technologies available, Lorber performed research on people suffering hydrocephalic disease (water on the brain). In his study, he found many of them had virtually no brain at all—simply a cranium full of cerebral-spinal fluid. Given that many of these people had advanced degrees or held high positions in organizations and all appeared as capable as those not suffering from the condition, Lorber asked this potentially disruptive question: is the brain really necessary?

Bear in mind that a normal forebrain's neural structure makes up about 80 percent of the ordinary brain, whereas the earliest hindbrain is quite small by comparison. Also note that the hindbrain, which includes the spinal column and peripheral nervous system, is the one that gives us information on body and world and how one survives in the other. Implications are that in Lorber's examples, the forebrain, which includes both the neocortex, or "thinking brain," and the higher parts of the emotional brain, were largely missing, whereas the primary sensory-motor and limbic system were probably still intact.

Consider further that mind and the field of mind with which the former is a reciprocal unit may not be just the outcome of normal development, as we assume. Mind might be extant from birth or even conception long before it has a mode or vehicle for its being. As shown in Lorber's cases, mind might also be fully functional in its relations with the hypothetical mind at large, making it possible for nature to compensate for the missing neural structures fairly well, as long as that old sensory-motor unit is intact.

We might, then, need to distinguish between parts of the brain missing from birth, when some compensation by nature might take place early on, and parts later damaged and impaired functionally. It is interesting to note that while a great deal of neuroscience has been gained by studying damaged brains, Lorber's findings would probably not be vigorously pursued because they indicate a top-down model eschewed by the academic world. Nevertheless, the notion of an original mind existing before birth or even conception is an important thought to consider.

All of this suggests that mind has a two-fold aspect. It may be, as proposed, an emergent of body-brain-heart interaction, but it also might be an emergent property of consciousness itself. This would indicate a universal field effect of mind potential that lies outside of or beyond any individual expression or employment of it—a kind of mind at large.

That mind may be an emergent property of a bottom-up brain on the one hand and an emergent property of a top-down field of potential on the other points toward the strange-loop effect discussed at the beginning of this book. It is this dynamic that may draw us to that which lies behind, beyond, or over and above all creative process and mental function, that Ground of Being, as termed by Paul Tillich, which seems the property of theological thought. Yet even beyond this lies a Vastness of which we are at times intuitively aware, though it is apparently even more unavailable to us than the Ground of Being. We can only *be* that Ground, whereas the Ground itself might arise from that Vastness that lies beyond all.

This elusive and subtle notion of a mind at large suggests a cosmology of interdependent creative systems in which, according to established theory, the wave gives rise to the particle, which gives rise to its wave. As with mind and brain, or individual and society, each is equally necessary to the other; each gives rise to the other or gives the other being. Meister Eckhart's provocative statement "Without me, God is not" does not claim that Eckhart thought he was God. The same qualification underlies all looplike polarities such as cause and effect, form and content, mind and brain, and consciousness and matter. Without the other, each is not.

Perhaps personal mind (Me) is rather like an intermediary between a universal potential, subtle and nonmaterial, and a material, organic function that realizes or expresses individual aspects of that universal. Able to move either way, drawing on and using either mind as emergent of brain or as reflector of mind at large, we are subject to both, for all in our awareness and worldview is brought about by this creative interplay. Here we find, then, a trinity of cosmology: heart, brain-mind, and mind at large.

LOST DIMENSION

Our modern, civilized consciousness has lost direct awareness of the heart-prefrontal dialogue. Certainly, most of the time most of us are unaware of warning signals the heart sends; thus negative events must slap us in the face for cognition, and this unbroken state of ignorance keeps us in a constant semialert mode against a world we can't trust. We know of aboriginal cultures that are not so cut off from intuitive heart knowing, though in each case their knowing acts selectively, according to the general history of that particular society and its people.

A cosmology referred to as Kashmir Shaivism, perfected before the tenth century, proposed that the primal creative power, which they named Shiva, a male energy in that culture's mythology, dwelt in the center, or "cave," of the heart. This nonmoving point was the silent witness to a female energy, Shakti, which issued forth from that hypothetical point as waves of energy spinning out and around us. These "wave forms of Shiva" were considered an energy matrix, like a womb containing the potential of all possible universes and creations. Out of this cocoonlike matrix, issuing from our heart and surrounding us in a circular swirl of energy, the creative Shakti spins out worlds for the silent Shiva within to witness. Without the center, Shiva, there is no Shakti; without Shakti, no

Shiva—and we are the field on which this play of consciousness and its strange loops take place.

In the early phase of my Siddha yoga adventure with my Indian meditation teacher Baba Muktananda, he spoke time and again of these wave fields emanating from the heart, surrounding us with love, power, and the potentials of all conceivable experience. I determined to personally experience that field at all costs: If it is present, I thought, surely we can somehow sense it. Through one evening and on into the night my meditation focused on that proposed field of wave energy. I put my strength into the "breath of fire" or *pranayama,* a deep and rapid breathing exercise that overoxidizes our body and thus, apparently, occasionally alters our ordinary awareness.

After an indeterminable period (seeming hours) of this intense, rapid breathing everything simply stopped—mind and breath suspended—leaving me in a state of utter simplicity, stillness, and clarity. It was then that I experienced, all too briefly, that electromagnetic torus field of the heart engulfing me, much as Muktananda described—as a cocoon of love and universal potential that I then spontaneously termed *plasma.* At the time I had no notion what plasma issuing from the heart might exactly mean but have since learned that *plasma* is a word for the huge "rivers" of electromagnetic energy pouring out of the sun, precisely as they pour out of our heart in miniature torus form. Thus we can see that this creative function within us is a condensation or contraction of a universal process that gives rise to our cosmos (see fig. 5.4). The scientific term for this organization of electromagnetic energy, *torus,* refers to a structure with a dipole or axis. A significant characteristic of a torus is its *holonomic* nature: Could any part be isolated (which, of course, is impossible), it would contain the information of the whole. Also significant is that all torus fields interact in hierarchies, so that the information in one field is at some point available to all aspects of the hierarchy.

Later, in April 1981, here at home in Virginia, I again experienced that wave field, this time physically and far more tangibly as a swirling energy and power surrounding and shielding my wife as she was giving

Figure 5.4. The top figure depicts a computer simulation of the torus arising from our heart, forming a dipole that centers from the pelvic floor to the top of the skull and then extends out in three distinct fields, likely the physical, emotional, and universal aspects of our heart torus. The bottom figure is an actual electromagnetic image of the torus from a living human heart in its first phase of formation immediately around the heart.

birth to our daughter, Shakti (so named by Baba Muktananda), in the early hours of morning.

In the late 1980s, at our ashram in India, an acquaintance of mine, Paul Ortega-Muller, professor of philosophy and religion at the University of Michigan, witnessed another aspect of this universal energy. Sitting in a dark hall for the thirty-day meditation retreat called the Blue Pearl, Ortega-Muller visually and kinesthetically witnessed again and again myriad wave forms of Shiva appearing directly in front of and around him. They were presented to him in full, three-dimensional color, and during every break, elated and excited, he rushed up to his room to sketch out his latest visions. For twenty years he had studied Kashmir Shaivism, which he thought to be an archaic, dead religion or ancient cosmology. Hearing that Siddha yoga was founded on this ancient theory, he had come to the ashram and enrolled in the thirty-day course. To have what he thought a dead religion brought to life in so dramatic a form as a living process or function within and around him was a life-changing event for him. He eventually completed an impressive volume of these wave forms and their significance. There seemed to me to be a rough resemblance between some of Paul's sketches and the color-coded renditions of magnetic-resonant images of living brains that have since been made. For years the Institute of HeartMath has researched the electromagnetic wave forms unfolding from our heart. It seems these fields arise from the Ground of Being itself, shifting and changing instant by instant, though they were presented to Paul selectively, according to his cognitive-formative structures of brain-mind and as reflections of what he truly desired and was capable of seeing.

Rudolf Steiner pointed out nearly a century ago, as did his soul mentor Goethe a century before Steiner, that the heart is not merely a mechanical pump but something profoundly more. "[T]he heart," wrote Steiner, "is the infinite circle contracted to a point. . . . [T]he whole world is within our heart . . . the complete inverted image of the universe, contracted and synthesized." Steiner also speaks of our world "radiating out from the heart," as the heart "senses its world." Steiner lived nearly a century ahead of neurocardiology and nearly ten centuries after those Shaivite scholars in Kashmir.

As I explained in *The Biology of Transcendence,* the first wave form grouping of the heart torus is apparently physical, the second relational-emotional, and the third our universal hookup—our heart's link to

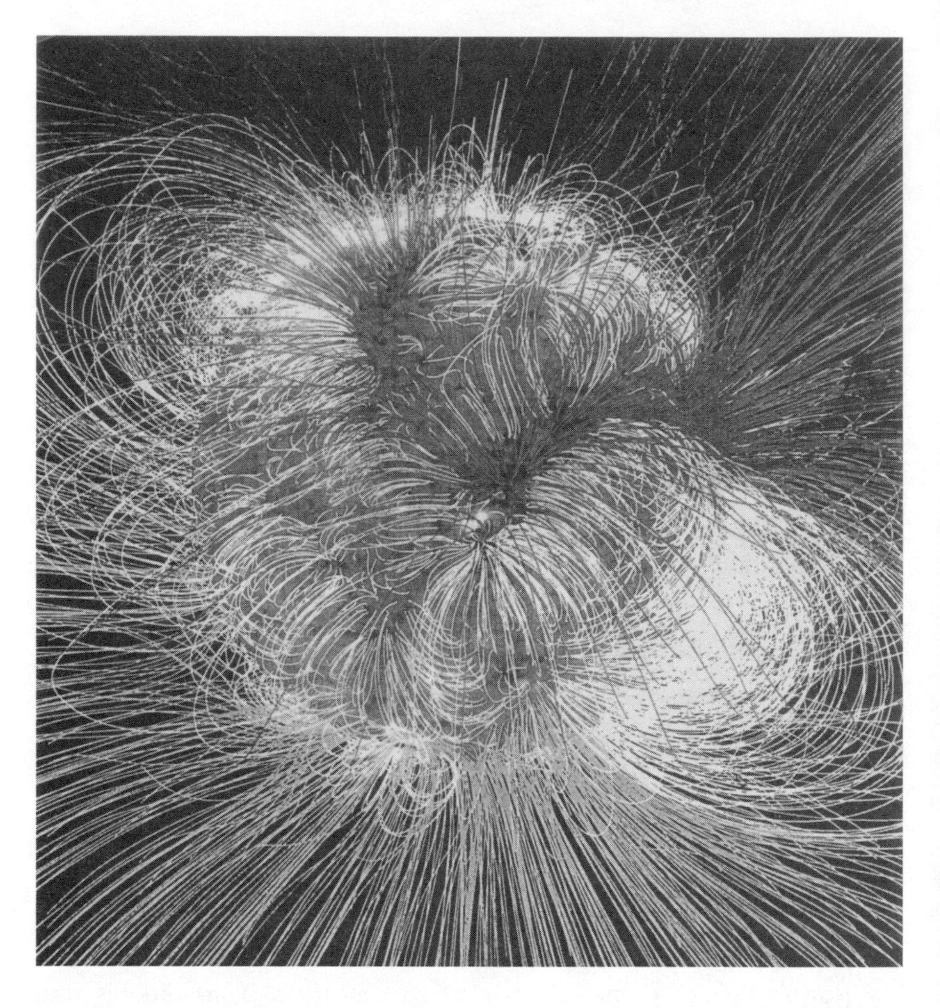

Figure 5.5. A depiction of a single instant's magnetic image of the sun, showing the multiple torus forms and rays of plasma moving out to encompass the solar system.

the nested hierarchy of electromagnetic fields making up our universe (see fig. 5.5). Again, all torus expressions of this sort are holonomic (any part contains information of the whole). Through our heart, we each are the center of our universe, and we selectively access that infinity as limited and shaped by our personal cognitive system, which is itself shaped by our experience of that which our species, culture, and personal models have selectively drawn and passed on.

As for the prickly question of a source of the heart's predictive knowledge, back in the lab at the Institute of HeartMath, we can see that there are formative orders of energy within the heart's hierarchal links by which it registers events that are still in the implicate or subtle formative orders of energy. This field of potential gives rise to explicate or physically explicit forms, to use David Bohm's schematic.

My own conviction is that the heart itself is the source of intuition, the nexus or meeting point of this interplay, as well as the source of both cause and effect and all such loops—a top-down interplay of major, cosmic proportions.

Thus the heart would be both the source of the information it processes through the brain and the source of that which it broadcasts to its body and world. As Rudolf Steiner explains, instant by instant the heart registers and reflects back the ongoing events of that world. Without the heart as source there would be no broadcast, and so no content. Without the expression of content, there would be no source. Without that which is created resonating out and reflecting back in, there is no creation. Without creation there is no Me or creator—and around we go, in wondrous cyclic loops.

6

PENFIELD AND STEINER

Nearly a century ago, Rudolf Steiner developed access to a field of mind he called the "higher worlds," a state that includes the phenomenon long hypothesized as the Akashic record, a kind of library or memory bank containing all the categories of human experience and knowledge, a morphic field including or included by this hypothetical mind at large. (In his book *Science and the Akashic Field,* Ervin Laszlo closely examines this Akashic field long referred to in more occult traditions.)

Through his higher-world connection, Steiner gave us a wealth of insights and a library of solid information covering a diverse range of topics beyond the scope of most educated people of his time. The areas he explored and expanded on were generally suggested to him by those concerned with current issues or technical problems. The answers or solutions he offered have proved continually more prescient as the twentieth century and now the twenty-first have unfolded. Steiner's life work came out of a new way of thinking, and we are often required to think in a new way to understand or apply his work. His often lengthy and detailed explanations came in response to questions from his students and others and were presented orally. He generally illustrated on the blackboard as he worked out his perceptions and the translations of insights he gained. Some of these explanations and concepts were written down by his students and he himself worked out and wrote down others. The vast scope and depth of Steiner's creative mind provides an intriguing history of its own.

Steiner gave us *anthroposophy*, a human-centered psychology, physiology, and cosmology-spirituality freed from the age-old prism of theology through which we had long looked at our reality and seen ourselves. From him came anthroposophic medicine, as distinct from the chemical-allopathic medical notions sweeping Europe at that time (and that have largely ruled our lives ever since). Waldorf education was his invaluable contribution to the development and well-being of children, in clear distinction from the mechanical, production-line theories of education arising from the ideas of Wilhelm Wundt, precursors to behaviorism, and others popular in his day. Steiner also gave us biodynamic gardening and farming, so opposite the chemical agriculture that was then appearing on the scene. He formulated an architecture based on avoidance of all right angles, which led to constructions of remarkable living shape, along with a theory of color based on our physiological-psychological-emotional interactions with our immediate surroundings. And he also gave us an invaluable knowledge of the heart that has slowly worked its way into our current research and understanding.

HIGHER FIELDS: OUR UNIVERSAL MEMORY BANK

Steiner's conception of higher worlds is both a product and catalog of the whole history of life on this planet and the entire cosmological process. Minds over untold millennia have fed into these higher worlds and continue to feed into them today, resulting in the mind at large. The various field effects discussed in the last chapter are integral parts of these higher worlds and its records, though it is important to note that the higher worlds are not a product storehouse managed by some "heavenly" intelligent guardian at its gates, nor are they governed by some cosmic true-false criteria that determines access to them. Feeding into this mind field are all actions, whether positive, negative, or neutral, creating fields within fields that both bring about the higher worlds and sustain them as they in turn sustain us.

From this universal memory bank or record file we can draw an endless laundry list of thought, most of which has been contemptuously dumped into the "idiot fringe" category by academic science. Certainly, some "fringe" nonsense muddies the waters of all thought, including science, and discrimination is necessary. Nevertheless, the academic world has tossed out legions of babies with the occasional bit of muddy bathwater. Here in this universal mind field akin to the Akashic record we find the origins of some, perhaps all, top-down phenomena. Here is the source of Eureka! experiences; the various forms of savant phenomena; genius in all fields; the source of the family of innate intelligences outlined by Howard Gardner; the frequency spectrum suggested by Karl Pribram as the source from which our brain draws and selects; the spirits that can "breathe us," as related by Eugen Herrigel in *Zen in the Art of Archery;* Gerald Feinberg's report of Kataragama "breathing" those initiates in his temple on Ceylon; and that "spirit of wholeness" to which Jesus urged us to be open, moment by moment, and to which he pointed when he claimed that "what we loose on earth is loosed in heaven." It is the reciprocal action between Steiner's higher worlds and our ordinary brain-mind-body experience that forms the relevant big picture in our discussion—and it is for us to avoid trying to localize the source of this top-down effect or, worse, sanctify it and make a god of it, which would build yet another tired religion from its endless potential. How our world forms moment by moment is through this reciprocal action between a top-down potential and our bottom-up action that manifests it, and whether we are aware of it or not, we are automatically caught up in such reciprocation: We can bring some nonexistent event into being through our passionate pursuit of it, as Laski's Eureka! phenomena show. The search for an answer enters into the creation of that answer.

Should we want to enter into this dynamic more consciously, however, then we need to consider the resonance between query and probe, question and answer, passion and search, and the fact that we find what we seek, though seldom directly. The system is too stochastic for any kind of mechanism to be attributed to it. Jesus, that cosmological genius, pointed out "Seek and you shall find," a fact regardless of the nature of either that sought or that found. Offending all equally, he noted that if you believe you receive, you will receive—positive or negative. In field effect there is no product evaluation, only process alignment. Reaping as we sow is not a mechanical karma, as it's often depicted in the East, but part of the strange-loop effect in our general cosmology. In fact, that sown in one generation might not be reaped until way down the line. Cosmological principals are amoral and beyond emotions, judgments, or demand for justice. Do away with judgment, in fact, and any need for justice disappears with it.

The Akashic or morphic fields Steiner investigated are an endless goldmine (and perhaps an occasional garbage heap) that we both bring about and have tapped into from a variety of disciplines, paths, and theories for centuries if not ages. Steiner's Waldorf education and his book *Knowledge of the Higher World and How to Attain It* were parts of his design to equip us to interact dynamically with these fields. Through the appropriate training, we can move beyond our own history of experience and even that of others' previous experience. Not to be dismissed is the possibility that fields have internal organization so that the potential of a field is continually reshaped according to new data coming into that field through general experience. The struggle Steiner saw generating around forces of darkness and forces of light, which culminated in the death of Jesus on Golgotha and which crops up again and again in history, is the source of the very "dark demonic" Carl Jung claims boiled beneath the conscious awareness of nineteenth-century Europe and led to both World Wars, and that today, sadly, rears its ugly head as vigorously as ever.

OTHERS WHO HAVE ACCESSED HIGHER WORLDS

Edgar Cayce, an early twentieth century psychic in the traditional sense, could draw information in a manner roughly similar to Steiner.

Yet he could tap into these fields only in a somnambulistic, trancelike state, a limitation that opened Cayce to charlatans who cashed in on his naive lack of discrimination while in trance. Despite those around him who may have been suspect, Cayce was a man of impeccable honesty and integrity, and even a cursory glance at the records show many phenomena he brought about that can't be accounted for with a classical bottom-up explanation.

Even earlier, Phineas Quimby, a shoemaker in Portland, Maine, whose psychic heyday came in the 1840s and 1850s, evolved a philosophy-cosmology based not on theory but on his own remarkable abilities and experiences. Quimby discovered by accident that he could find people who were lost, and he developed a keen intuitive and even precognitive sense. He detested doctors and instead intuitively determined the emotional nature behind most diseases. He would locate himself in close proximity to an ill person and on some subtle level fuse with the individual's psyche or emotional state, identifying with, resonating with—in effect, becoming—the illness from which they were suffering. He would then return to his own body awareness and throw off the disease, having determined that it was largely caused by an emotional upheaval or perhaps an invasion by a malevolent influence, at which point the ill person's sickness reportedly vanished with it. He was, of course, swamped with demands for his services, and he hired a secretary, Mary Baker Eddie, to assist him in keeping his records and managing his copious writings in which he explained his point of view. This secretary was purported to have purloined his notes, from which she created her own system, Christian Science. Quimby's heirs objected (or wanted a slice of the proceeds) and lawsuits concerning the issue ran well into the twentieth century.

Quimby was preceded by the eighteenth century's Emmanuel Swedenborg, from whose mind-set and writings a bona fide church-building, Bible-thumping religion formed. Swedenborg's history is a fascinating story: He was the most famous scientist and engineer of his time until, in his midfifties, when he had a mystical metanoia or mind shift that opened him to aspects of the higher worlds and nonordinary modes of thought and experience and a new aspect of theology. But William

Blake scorned his theology, claiming there was more in one line of Jacob Boehme than in all Swedenborg's efforts together.

People of such capacities have appeared throughout history, giving a field nucleus to that movement growing throughout the nineteenth century concerning what we would call psychic phenomena. This basis gave rise to the more sophisticated "new thought" school in England in the late nineteenth century and the whole "power of positive thinking" school prevalent in the first half of the twentieth century. The strange phenomenon of a Jewish woman in New York being seized by a spirit presenting itself as Jesus and speaking to her through automatic writing brought about "A Course of Miracles," a "self-study spiritual thought system," which was accepted in the late twentieth century and is still religiously followed by countless people, including university professors, medical doctors, psychologists, and an illustrious segment of our intelligentsia.

William Blake said, "I must create a system of my own or be enslaved by another man's," which seems to be a universal impulse and a safe one to follow, provided we can resist the temptation to gather a following of our private creation. In addition to offering solid cosmological principals, Blake, shoemaker-philosopher Quimby, and the rest were characteristic of individual minds breaking out of a cultural shell, although that shell immediately repairs itself in those broken spots.

Suspending Cause and Effect

In *The Biology of Transcendence* I related how, as a young man of twenty-three, I discovered by apparent accident a state in which ordinary cause and effect could be suspended. Here is a condensed version of one of my experiences related to our question here: what is mind?

In 1950, I worked the graveyard shift in a bank clearinghouse and attended university most of the day and was thus seriously sleep-deprived. To catch a few winks, I discovered a way to run the IBM check-proofing machine I labored with (which normally involved a fairly complex, high-speed, two-handed, wide-awake skill) and sleep at the same time. I had already discovered a state I later termed *unconflicted behavior,* in which ordinary cause-effect could be bypassed. By trusting without doubt (not as simple as it sounds and yet effortless), I discovered that some other part of me could take over my job and allow my ordinary self to sleep. I even dreamed!

My method, it turned out, was so efficient that I outperformed the other operators on that midnight shift, never making an error and increasing my output to about fourteen thousand items per shift. All this and I could stay awake in class the next day. After weeks of this saving grace, my supervisor discovered what I was doing and on threat of immediate dismissal, I had to abandon the practice. In order to maintain the state, I had to skip many of the rigid bank rules for regular checks and balances throughout the night. Pausing frequently to verify my accuracy in this way interrupted my sleep and would, I suspect, have indicated my doubt that this other part of me could do its job. If an error *had* occurred, however, it could have taken all day to go through those fourteen thousand items to trace the mistake. My supervisor could in no way risk this. And so, while I knew with absolute certainty no error would occur, I faltered in my explanation of my ability and my faith in its fail-safe accuracy was not shared by the nonbeliever in charge.

In retrospect, the enigma brings us to ask who was sleeping and who was awake. Which Me was connected with mind? If both—what is mind that it can function in two opposite modes at the same time, as suggested by my experience and that of conductor Artur Rodzinsky and my brother's acquaintance in the signal corps and, as we shall see, the famous operations of Wilder Penfield concerning target cells and music memory? I had also learned that the will of this elusive "It breathing me" was able to bulldoze through and overwhelm the will of people in ordinary conflicted behavior. I began to realize that this capacity could easily slip into the demonic mode, and though I knew the capacity itself was amoral, I eventually gave up on pursuing it.

Still, the experience of functioning in two modes at once, both operating the IBM machine and sleeping, has remained puzzling to me. I have rarely experienced such paranormal phenomena since. Perhaps, paraphrasing Wordsworth's language, the world has been too much with me.

Further, as Blake suggested, he who doesn't believe in miracles makes certain he will never take part in one, and my belief in myself has eroded. Years later, Baba Muktananda echoed his predecessor from a couple of millennia earlier: "Doubt is the enemy of spirit." I began to suspect that this spirit was what ran that IBM machine without operating according to any of my criteria or that of my culture.

Meanwhile, evidence has been growing for a top-down cosmology in reciprocation with bottom-up functions. Mind can interact with matter on the one hand, and mind at large on the other, thereby setting up strange loops of causal effects not available in our ordinary dynamics.

Penfield's Mystery

In his electrode brain probes of some fifteen hundred patients over the years, Wilder Penfield mapped out the surface territory of the neocortex. Because the brain has no feeling and only a local anesthetic was needed to remove the top portion of skull, Penfield's patients were fully awake and conversed with the doctor, describing their experience as he probed their brains with electrodes.

Time and again, Penfield found certain "target cells," which, when activated by an electrode, brought into the patient's conscious awareness a complete replay of some past event in that individual's life. When such a target cell was electrically activated, the patient gave a play-by-play account of what he or she experienced. While the patient could describe in graphic detail the sensory experience he or she was undergoing, at the same time he or she recognized the experience as a memory from childhood or some event long past. These events were not similar to Eureka! moments, happening all at once, but instead unfolded in ordinary time as in the original event, during which the patient felt directly involved in as full a sensory level as originally experienced.

Often, Penfield ran across a target cell that activated the patient's hearing a piece of music, most often from childhood, as distinctly as in a live performance. This too, as with any of the other target cell memories, would continue as long as the electrode was firing or until the memory episode had run its course. Should the electrode be left in position on the target cell, upon the doctor's restimulating the cell, the event would replay in its entirety. Because the patients were looking at Penfield as they gave these reports and responded intelligently to Penfield's queries, each time Penfield and his patients wondered how two simultaneous events could take place, one that happened years ago and the other happening in the present in the operating room conversation with Penfield. All of this Penfield discussed in his last book, *The Mystery of the Mind,* and he had the grace not to attempt to explain away any of it, admitting it to be a mystery to him.

Penfield's nurse assistant, who had been at his side through all those years of this pioneering neuroscience, was a frequent visitor to our meditation ashram in India, and we had long discussions about Penfield, his methods, and his general record. She spoke of his extreme care in every move—an operation might take many hours—and the meticulous notes he made of each of those fifteen hundred surgeries. In most textbooks on Penfield, reference is made to his mapping of the "homunculus," the surface area of the neocortex involved in body sensation, but discussions of the more esoteric findings that were of the most interest to him, events that figure so largely in his notes, have been generally ignored.

As a footnote to Penfield's intriguing work, the proposal was made that certain "grandmother cells" in the brain could hold an entire memory. This idea was qualified when it was discovered that such cells, when activated, in turn activate many areas of the brain involved in the remembering, or piecing together, of memories. This process may well involve grandmother cells as part of the greater web involved in memory and possibly learning, and the theory of grandmother cells continues to be researched and enlarged upon.

Nobelist Gerald Eddleman argues that memory may be continually updated by our brain and is, at best, often quite unreliable. Many a good man has been hanged on the evidence of someone's tenuous and questionable memory. We tend to update memories to fit new understanding or knowledge, to maintain our integrity or ideation, or to improve our image of ourselves, and once we've remembered in some novel way, we protect that novelty fiercely because often our integrity gets tangled in the process. The mind, it seems, can bend back on its own brain and shuffle around materials, just as brain might shuffle around mind.

In this chapter we have only touched the tip of the enormous amount of information concerning anomalies of brain function, such as the astonishing work at the Princeton University's Engineering Anomalies Research Laboratory and the work of Brenda Dunne and Robert Jahn. No matter how great the mass of research, conjectures, and speculations, however, the mystery of mind and its brain remain just as mysterious as ever.

Part Two

The Conflict of Biology and Culture

INTRODUCTION TO PART TWO

A few decades back, a young anthropologist, Margaret Mead, paid lengthy visits to a number of Polynesian societies in the South Pacific. While a decade earlier anthropologist Bronislaw Malinowski had reported at length and far more graphically on the "sexual behavior" among the people of the Trobriand Islands, a storm of controversy arose on publication of Mead's reports that these Polynesians imposed no sexual restrictions on their children or adolescents. Each village of the Trobriand Islanders, for instance, had a "children's house" in which pubescent children congregated, exploring sexuality in all its wonder and excitement.

Attractions and affairs must have cropped up in a constant turnover among these Polynesian adolescents, couples forming and parting, loves ebbing and flowing and waxing and waning, each youngster trying out again and again his or her sexual-emotional fit. After about four years of such experimentation, each young person began to settle down with that one partner whose resonant frequencies were most lasting. These maturing couples then separated from the group, went through whatever ceremonies their society called for, and began life together. After a time, when they decided to have a child, the prospective mother asked permission from her guardian goddess and soon became pregnant.

The intriguing part of Mead's reporting was that in this entire four- to five-year period of intense sexual activity, no pregnancies occurred and apparently no cloud of such expectation or concern had ever fallen on the sexual sorting and selecting taking place. Mead wrote of the "adolescent sterility" that had prevailed. It seemed as if this protective stance was the cultural expectation, and cultural expectations tend to be lived out by cultural members. It almost goes without saying that similar behavior would not be acceptable in our culture, which largely determines our mind-set; nor, because of this, would such sterility hold.

On publication of her work, a goodly portion of our populace was aghast. Mead herself was sharply criticized, and more than one critic set out to discredit her. Her reports on sexuality were largely buried, in fact, much like Darwin's *The Descent of Man.* In the late 1980s an anthropologist followed Mead's earlier travels, interviewing the few elders remaining who could recall her visits, including some who had even been interviewed by her. He found no traces of the open promiscuity Mead had witnessed and subsequently made quite a fuss over his findings. These elders reportedly had not only denied such youthful laxity but seemed embarrassed about the earlier accounts and claimed nothing like that had ever taken place.

Eventually, other anthropologists noted, in rebuttals of this rebuttal, that in the several generations between young Mead's visits and the later visits by an anthropologist who openly set out to discredit Mead much had taken place, notably the Christianization of the Polynesian people. Even those resisting outright conversion had been equally caught up in the heavy atmosphere of guilt, sin, and promised punishment should their sexual license continue, and to be sure it had not been continued or was sharply curtailed and hidden. Guilt is a powerful tool and can dowse the fire of passion against considerable odds. Furthermore, through the "molecules of emotion" brought on by the ensuing guilt, even if intercourse and pregnancy occur in spite of injunctions against it, the negative emotions accompanying such occurrences will take their toll, as will be explored later here in part 2. Interestingly, UCLA's Patricia Greenfield writes that "mirror neurons absorb culture directly," while other studies show that "social emotions" such as guilt, shame, pride, embarrassment, disgust, and lust are automatically imprinted by mirror neurons and fire automatically, below conscious awareness or intent.

By the time of the later studies, clothing in these Polynesian cultures was not merely for adornment but instead was used to cover shameful bodies. Many facets of Polynesian culture had shifted as a result of the new theology, and indeed the neuroses, angers, anxieties, and problems of modern Western life had dutifully moved in to fill any gaps. Nevertheless, the people were being good and behaving morally, as Christians should, though the old courtesies and benevolence were fading as was the constant happiness that used to prevail.

The point of this account is nature's clear Darwinian selectivity in those original people. That four-year trial period among the group's youth was nature's way of seeking out the best of all possible DNA matching, a basis for a stable family life and society into which new life could come. These people sought the resonant love that builds bonds without bondage, and it just might be that nature knew what she was doing in this profusion of selective screening. Is it simply coincidence that such preliterate societies generally lived in a balance of population to environment, with mass overpopulation seldom occurring?

Nature operates by a model of profusion, with masses of potential selectively giving rise to singular successful forms and massive chaos giving rise to specific order. The form of the order is the result of a stunning intelligence, but that intelligence is the result of the movement from chaos into order. Whatever ordering movement is eventually worked out is equally the intelligence working out that order. To assume, as a fundamentalist would, that the intelligence precedes the order is to miss the wonder and mystery of the looplike process of creation itself. What lies beyond or comes before this process is the Ground of Being beyond which is the Vastness—and, as the Rig Veda asks, who here can say from where that comes?

Human females are born with somewhere between four and seven million eggs on their ovaries. Were all the eggs of a common housefly to hatch and reproduce for a few fly generations, within a year flies would cover the earth in a layer many feet thick. Likewise, were all a single human female's eggs to result in embodied humans, the earth would shortly be radically overpopulated. Between birth and puberty, however, nature selectively whittles away at this mass overproduction of eggs until, by the time of menarche, the number is down to a mere four hundred thousand or so eggs. What determines this reduction?

Each egg potential is a particular folding of DNA, no two of which are identical. DNA is sensitive to a wide variety of signals from mother, family, society, and world. The overall frequency or resonance between the world out there and the egg cluster within undergoes constant change, during which those egg potentials with less resonance are eliminated. With an available four to seven million variations on the general theme of human life, just about any possible contingency of change can be covered. After fourteen or so years of this not-so-random selectivity, the egg survivors are down to a small percentage of the original mass.

At the onset of menarche, that category of DNA that has the closest match to the current signal input selects as an ever-changing group—ladies in waiting, we might say—for the further sequential selection of the upcoming bride of the lunar month. For this, months of preparation are involved. That bridal egg must be brought to maturity, encased in its bridal adornment of a protective shell (the cellular membrane so critical to life, as Bruce Lipton made clear—hardly like the shell of a hen's egg that we crack at breakfast), and finally ushered to the fallopian tube to be dropped down into its final position at the altar, ready to receive its suitors.

Our male children, on the other hand, are born with no active reproductive functions or suitors yet to come. Such creative work will not even begin until that period of puberty, just about the time those lovely ladies are preparing for their monthly offering. Once that male production line begins, however, it is an ongoing, nonstop, twenty-four-hour-a-day operation. By this delayed and then constant production method, each male sperm is automatically updated moment by moment by that male's environment and his own experience. Theoretically, by maturity, he can contribute enough of those aggressive little creatures in a single encounter to fertilize every fertile female on earth who might be ready for fertilization at that particular time.

At each union of male and female, three to four hundred million of these competitive little knights charge in to swim madly upstream toward that lady in waiting. This is a staggering, massive number considering that only one sperm can finish the venture, and even then only should the best-case scenario unfold, for an equally massive on-the-spot selectivity takes place in the last upstream leg: the river itself weeds out all those candidates who just don't measure up.

By the time the egg is actually approached, the number of suitors is few indeed, and even this paltry few must go through a final selection. These remainders reportedly form a ring around that egg, which is many thousands of times larger than the tiny suitors who, in effect, are parading at the court wherein a final frequency-matching between egg and suitors apparently takes place to determine which lone little sperm is invited into that inner sanctum.

Think how many millions of possibilities have been selectively screened before the best possible match is found and the egg finally throws open her portal—a gateway just large enough to allow entrance of her chosen spouse (at which point the portal quickly closes, for competition in that sanctuary would hardly be appropriate). That long, thrashing tail that propelled our hero through its journey now ceases oscillating and disappears and the remaining DNA bundle is propelled to its female counterpart and destiny.

Years ago, a biologist proposed why females are born with eggs-in-waiting while males must spend several years waiting just to enter the game: The female's egg holds the template for the affairs of state at the time of her conception, thus providing a stable base for the genetic system, and about twelve to fourteen years later, that stable base will be unchanged, even as selectivity according to environment and DNA folding will have occurred. The male's contribution will be a variable, an update factor determined by the changing environment and the male's own ever-changing experience. The stable biological base of the female without the variable of the male would lead to stasis, with adaptability and evolution less possible. The variable of the male's sperm introduces instability, but also adaptability—that is, an ability to change as life itself changes. If there were two such wild-card variables, the process would be chaos, while two stable bases would produce stasis. One of each and you have us.

As biologist-anthropologist Ashley Montague pointed out, the female, as the stable biological base of life, is the stronger of the two needed to create life, while the male variable is more fragile. Note that in reproduction a single egg awaits while hundreds of millions of variable males run to meet this one. Being a male is precarious: a sperm is more difficult to produce and keep alive and, sadly, is expendable after contributing his small part.

In the several natural spontaneous abortion periods of fetal life (such as around the tenth week and the fifth and seventh months) 80 percent of the aborted embryos or fetal infants would have been male had they not been aborted—and odds are strong that they would have been dysfunctional if they had been carried to term. Of all dysfunctional children born—those who are deaf, blind, missing various body parts, deformed, and so forth—a majority are males. Of all autistic children, a majority, again, are males. Adult females outnumber males, generally live far longer than males, and can live without males altogether far more successfully than males can live without females. Sigh! (As an aside, it is interesting to note that Ashley Montague's milestone work *The Natural Superiority of Women* was written long before the women's movement began.)

Strangely, in Western societies male infants, though more fragile, are given less care than females; they are "toughened up" according to our current mythos, apparently to deal with a hypothetical saber-tooth later on. Meanwhile, we overprotect our little girls. In peaceful, preliterate societies, however, male infants are breast-fed far longer than females and are generally treated as more fragile. It seems our mythos betrays us, however. Nature gave humans our higher brain structures that we might outwit saber-tooth, but she did not give us larger, tougher bodies to outwrestle him. We have seriously misread the signals. Without nurturing, our males grow up aggressive and angry, do not make stable families, and are prone to violence against spouses, offspring, and each other.

In their respective works on brain and development, James H. Austin and Allen Schore both point out that our survival today depends on producing males who are nurturing, benevolent, compassionate, and caring, which, in turn, requires just such treatment of males from birth and particularly at the toddler and adolescent periods. This notion, however, is so diametrically opposite current neo-Darwinian capitalistic cultures, with their emphasis on competition and survival of the fittest, that the chances of such a radical turnabout seem slim—as indeed our general survival itself does—but we might at least give it a try.

So life and its creation is stochastic, with random chance a factor, but an obvious purpose lies behind this randomness and the nature of the selectivity reveals an intelligence of profound depths. In fact, the whole scope of creation reveals a strange-loop design of incalculable brilliance, as Walt Whitman pointed out in looking at a simple blade of grass.

7

NATURE'S BIOLOGICAL PLAN

New Zealand biochemist Michael Denton argues that the universe drives inexorably to produce life, which is to say life drives inexorably to express itself. Looking at the Hubble telescope's pictures of deep space, we see universes beyond universes unfolding, apparently forever, each consisting of billions of stars and their planets. If Denton is correct, all this grandiosity has the single purpose of producing life, and we—you and I—are the high point of life on this globe.

RANDOMNESS WITH PURPOSE

Academic science has been built on the dogmatic assertion that life is an accidental by-product of a random-chance event. This neo-Darwinian belief protects the sovereignty of the scientific priesthood and elimination of any top-down influence outside scientism's prediction and control. That our origin was by random chance is a half-truth that, like most half-truths, can be doubly deceptive but may lead to a whole truth if pursued.

As we have seen, life unfolds stochastically, randomly, and with purpose. To deny a random factor in our unfolding is blindness, while to ignore purpose is plain foolishness. Life is as high a purpose as we can conceive. Yet those who claim that everything happens according to plan or by intelligent design are equally blind. Accidents in creation happen all the time as each possibility for expressing life works out its accommodation with all other possibilities being expressed in a continual unfolding of possibilities. The one that works best persists as long as do its accommodations to the myriad other creations following the same schematic. Without this stochastic factor, both our biology and universe would have to be machines, which neither even remotely resembles.

The scale on which nature unfolds is infinitely huge when viewed through scientific technologies such as telescopes, magnetic imaging, electron microscopes, and so forth. The immensities found in both directions, within and without, seem to reduce us as individuals to utter insignificance, as is so often implied by scientism. The Greeks, by contrast, looked at the nameless stars and wrote their names all over them, and all was well.

But our implied insignificance can be tempered by two simple observations: First, we are as big to the smallest as we are small to the biggest. Modes of being are relative. Our body, for instance, is made of upwards of seventy-eight trillion cells, each a creature functioning semi-independently and intelligently, though by and large in harmony with its neighbors. If we look inside the inner cosmos of which we are made, we find it to be as bewilderingly immense as when we look out at Hubble's cosmos. Both directions require the same general type of scientific technology to observe, and though we could live here happily without either observation, we are curious by nature—nosy, inquisitive, attracted by anything new. Chimpanzees or bonobos, for instance, stay in the same locale, by and large, but apparently humans, from their beginning, wondered what might lie over that next hill, whether literally or figuratively, physically or mentally.

Discovering the cell was on the same order as discovering the moons of Jupiter and made for excitement and adventure for those interested in such discoveries. In time, our awareness of the cell's complexity increased with the complexity of microscopes used to travel there. The electron microscope revealed a whole world within each cell. In simplest terms, a cell is a body cradling a nucleus of DNA that folds and unfolds within it. The substance around that DNA contains tiny creatures called mitochondria, who busily convert energy into food and vice-versa. All this is in a filling or substance called *tubulin*, which makes up much of the cell's interior. Tubulin, in turn, is made of microtubules that are oscillations of frequencies, a force that turns off and on untold times per second. When on, the microtubules are there, when off, they aren't. This seeming on-off whimsy gives a solid basis for cellular structures, but if followed, undermines the one absolute eternal law of physics that has been writ into the stars and heavens: Energy can be neither created nor destroyed. But the final, ultimate truth is the fiction of a stuff called *energy.* If energy "turns off," and at that point is not present, where is it? It is no more here than when it is "turned on." The oscillations of the microtubules are the very genesis of the strange-loop phenomenon that gives rise to matter and the universe it can manifest. As Robert Sardello explains, energy is a proposal of early science that has never really held water but was invented as a semantic screen for an unknowable force beyond our conceptual frame. Thus, those ghostly, oscillating effects such as the neutrino and microtubules are labeled for convenience by science to screen out a category of being that would simply unhinge scientific tautologies if brought to light.

A sperm is a very tiny cell consisting of a body that is not much more than a folding of DNA and a tail or flagellum by which it propels itself toward the egg. That tail is made of nine microtubules arranged in a circle and supposedly powered by mitochondria. On admission to the egg, the tail falls off, as they say, but in actuality the mitochondria simply stop powering the microtubules, at which point they aren't—because they are not needed any more. So tubulin, made of countless microtubules and making up much of a cell and thus our body, is of the same order of magnitude in reverse as that cosmos out there (which we assume is made of more tangible stuff—though such stuff is still an assumption we need lest our mind-set truly fall into chaos. I refer you to Lynn Margulis and Dorian Sagan's monumental works in microbiology.).

The second observation tempering our insignificance is that when you own the mint, as nature does, you don't have to pinch pennies. Said differently, nature's profligate and endless redundancy and massive apparent wastefulness of so much to achieve so little can't be overstated. But why not operate this way? Create one particle or wave of energy, if only as an illusion or play of mind, and relate it to another—such as nine microtubules combined to make a perfectly serviceable tail—and the process is put in motion to create anything and everything forever.

This cyclic movement has no need ever to stop because in the strange-loop effect found everywhere in nature, creation doesn't begin anyway—at least not in any logical or linear way according to our sensibility. Creation simply *is,* an ever-present origin that hinges on the relationship of things that were themselves created out of waves, vibrations, or frequencies. These, like microtubules, can't be proved to exist except by use of a device made from such waves, vibrations, or frequencies—really, a kind of tautology itself: Waves, vibrations, or frequencies must exist to prove that they exist. (Interestingly, only technological scientific devices can prove the validity of modern technological science, which means that the sacrosanct scientific method is also a kind of tautology.)

Something that exists is outside of or set apart from. A wave frequency is set apart from only another wave or from itself. The early child, in its play, creates an interior imaginary world. He then projects this on the exterior world of his senses, which are assembled and sense by that same imaging brain-mind. He plays in what developmental psychologist Lev Vygotsky called the "modulated reality" of his own "let's pretend," but such play can never be other than actions of that very same brain-mind. This amounts to a true and functional tautology—a tautology that can work! All the child wants to do is play in this state, a kind of divine play in which he has dominion over his created world. He is actually playing God. Stop the child's play and you kill both his rising dominion and natural divinity.

So stochasm indicates purpose amid random chance. Technological science demands clarity of disciplined approach, economy of procedures, eradication of random chance if possible with little or no wasted motion, and a maximum illusion of predictability and control. If, however, the way by which stars, planets, moons, and space junk are created could be put under a single rubric of organization, it would be "order out of chaos," making chaos the first order of the day. Chaos indicates a realm of pure randomness, therefore endless potential and plenty of raw materials to order into being.

In nature's ensuing litany in which order out of chaos endlessly reproduces itself, as does the chaos needed for order to be, you have a production procedure wherein anything and everything that could ever possibly happen or be created has, is, and always will be. Thus purpose, life and its ever-emerging through individuals, will always be displayed along with all that randomness that just happens to be a critical part of the production process. So what if one galaxy collides with another, annihilating both, and that sort of thing occurs right and left? Even though that collision would probably take a billion of our years to be completed, there are untold millions of galaxies and untold years left and more are always in the making. So what if only one of a billion stars hits the jackpot and produces the right conditions for life to appear? In this nothing has been lost and everything gained. You and I are here and the universe is aware of itself anew and wants only to do it again. More! More!

In every field of random, purposeless, useless junk happening or chaos in this "stately dance to nowhere" (as Harvard's paleontologist and evolutionary biologist Stephen Jay Gould termed it) lies this pearl of great price—the very conditions from which life will spontaneously spring forth, thereby exposing Gould's error for what it is. The nowhere Gould referred to is actually everywhere and his stately dance is eternally underway, with the goal always attained, again and again, instant by instant. And pearls such as you and me will always be found, because fields of chaos keep forming to bring them about.

This justifies the apparently prodigious effort behind creation, making it all worthwhile. Yes, our cosmos is a stunning, magnificent, and overwhelming indulgence by that force behind it that apparently desires to have Being—and how else should a creative process behave?

Walt Whitman said the simple quahog in its shell was enough (that is, enough to prove the wonder of creation), and William Blake said, "More! More! Is the cry of a mistaken soul. Less than All will never satisfy." In our scientific view, this necessitates an expanding universe. Whitman was awestruck that something so uniquely complex and wondrous as a blade of grass—something he saw as a display of creation as great as "the journeywork of the stars"—should actually *be.* Any one of us, should the scales fall from our eyes, would be awestruck as well. Blake found a world in a grain of sand, eternity in an hour. Some poets have always seen.

No effort is actually involved in creation. With no effort, the child at play creates worlds to inhabit. Creation is free. Only when we arrive at higher, mature forms of life, such as adult humans, do we see that this free ticket is lost and charges for everything start accruing. We then tend to make effort of everything, perhaps as a means of establishing self-importance or giving meaning. This peculiar human compulsion taking the form of a search for meaning or the desire to be important arises from a notion that we are unworthy of such a grand event as this life, a peculiar pathology that seized us somewhere along the line of ancient history and wrecked our play. We even act as though we're not supposed to be here at all and must somehow justify our existence or buy our way with blood, sweat, and tears. Because of our enculturation, we can't grasp the idea that as the end-product of this stochastic trick of creation, we are, by default at least, co-owners of that process that brings us about. No matter at what remove, we are heirs to both the mint and the process of making mints, yet we seem to prefer to squander our lives in scrabbling, cheating, lying, stealing, murdering, and warring—over pennies.

BRAIN DEVELOPMENT AND THE RISE OF THE PREFRONTAL CORTEX

In his first major work, *Origin of Species,* Darwin showed how life's species arose through random-chance mutation and a survival selectivity. In his second, *The Descent of Man,* he described a moral imperative of love and altruism necessary for the human to have arisen and survive. As we have seen, nature couldn't have unfolded the second work until the first had prepared the way. Thus the two works, Darwin 1 and Darwin 2, present a strange-loop effect: each is found in the other and gives rise to the other.

Those involved in brain research at the Howard Hughes Medical Center have studied the DNA of our numerous predecessors and recently made the audacious claim that our human brain appeared in far too short a time to be explained by Darwin's process of random mutations and survival of the fittest. The human brain to which they refer is the fourth and latest brain added by evolution, the prefrontal cortex, which is made up of the largest lobes in our head and lifts us above all other animals. Yet this novel fourth brain rests on the shoulders of and is critically dependent on those older animal brains in our head—the reptilian and the old and new mammalian. Nature had spent eons bringing about these remarkable Darwin I brains through random mutation and survival selectivity, and in her economy, she wasn't about to reinvent those wheels when it came time for the next brain to be formed.

So when these first neurological evolutionary achievements had set the stage, nature brought us forth pronto. She had only to relocate a few physical parts, get rid of some outdated ones, and add a new brain that was a product of and in turn produced love and altruism—that is, love of self and others. This was not an overnight affair, but certainly involved less trial and error than was involved in the creation of those previous brains and the transition probably didn't take too long in evolutionary terms.

As we've learned, those lower animal brains are the neural systems that give us our world and body knowing, and these formed the basis for the new, precocious upstart, the prefrontal cortex. It was neuroscientist Paul MacLean who first articulated the evolutionary nature of these three brains. The basic reptilian sensory-motor brain interprets our body and environment. The old mammalian or emotional-cognitive brain brings onto the scene relationship with others, and a neocortex or new mammalian brain added atop the first two is critical to the precursors of speech and thought. These three grow their basic cellular structures in the mother's womb in the same sequence, though with overlaps in between each stage: reptilian first, then mammalian, and finally new mammalian. After birth from the womb, they develop again in the same sequence, more or less, with still more overlaps. (Refer to figure 5.3 on page 69.)

The fourth brain, the new prefrontal cortex, grows its cellular structure after birth and ends as the largest structure in the brain. It is so large that if it grew its structure in utero, we would never be able to be birthed. It appeared as recently as only forty thousand years ago in our history and thus has less evolutionary weight behind it than the older brains. While far more powerful when it is completely formed, this newcomer is quite fragile until fully matured, which takes roughly twenty-one years after its postbirth appearance. A rule-of-thumb is that the older the neural structure, the more durable, instinctive, reflexive, and nonnegotiable it is, since it has no capacity for language or reasoning; while the higher up we go on the evolutionary scale, the more fragile the system, the more time it takes to develop, the less instinctive and more negotiable it is, and the more powerful it is at completion.

Nature's plan for the development of each of these four structures follows the same process as their orderly evolutionary appearance. A similar evolutionary order can be seen in the germination, division, and growth of cellular life from egg to embryo to fetus to uterine infant to newborn. This schematic roughly recapitulates the development of all evolution's life forms that came before us, culminating in birth out of our first matrix, our mother's womb. The developmental stages, unfolding after birth, follow the same sequence, with much overlapping and looping back and forth between them (fig. 7.2).

At the medical school of UCLA, developmental psychologist Allan Schore, author of the monumental work *Affect Regulation and the Origin of Self,* calls this sequential unfolding the "evolutionary ladder." Each rung of the ladder represents the development of each of our animal inheritances in their respective order. We get to the second rung by developing the first, and we stand on it to move up to the third, and so on. The top of the ladder brings us to the development and reign of

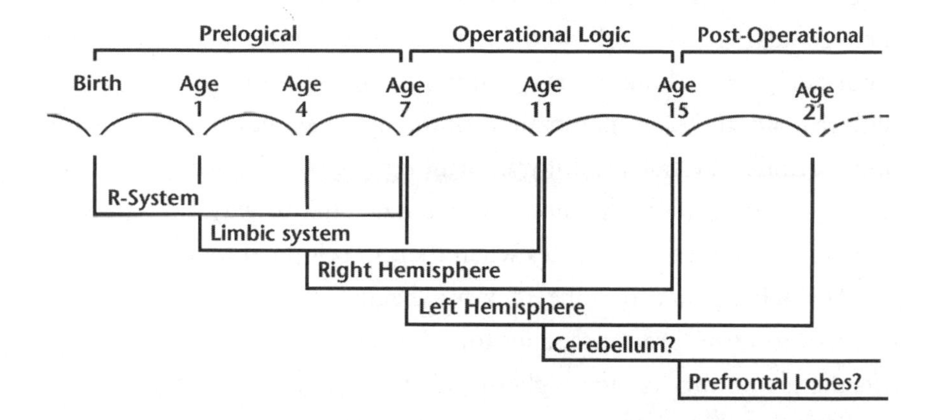

Figure 7.2. Brain growth spurts and shifts of concentration of development

the prefrontal cortex, that fourth brain added by evolution as not just icing on the cake, but the purpose of the whole show. Nonetheless, the actual growth and development of this highest fourth brain requires as its foundation the development of each of our animal inheritances in their respective order.

Although containing many different parts, our brain functions as an integrated unit, as neuropsychologist Karl Pribram pointed out years ago. Any activity may involve a particular module or lobe of brain that specializes in that particular action, but this automatically involves the rest of the brain peripherally. All parts act in a supporting fashion while at the same time registering or imprinting the general gist of the action. Through this, we have ongoing development and continuity, a brain-mind ever expanding yet maintaining its integrity. This proves a workable pattern for life in general, families, communities, musical ensembles, casts of theatrical plays, and so on. If we are totally caught up in our emotions, for instance, as in lovemaking, all the brain-body supports this great enterprise, it is hoped without any one part getting in the way of the other. If we are involved in intense intellectual activity, our forebrain dominates the scene, with the hindbrain taking a back seat, playing only physically supporting roles.

After birth, the modules and lobes in our skull continue the same sequence of development, which unfolds in much the same pattern that their cellular growth followed in utero. The reptilian sensory-motor hindbrain gets the bulk of attention in the nine months following birth, with the old and new mammalian brains lying semidormant, involved only peripherally but dutifully imprinting a general resonance of this sensory-motor activity that holds center stage. In this way, they are well-prepared to slip into their roles when their developmental turn arrives.

As each module or lobe of brain begins its higher development, incorporating its predecessor as foundation and auxiliary support, it lifts the lower into its own higher mode to some extent, completing the development of the lower as it does so. Thus there is a constant developmental overlap. Just as there are no straight lines in nature, brain development is a curvy, looped affair, less stringent and linear than figure 7.2 indicates.

In traversing this evolutionary ladder of development, windows of opportunity open and close for us according to precedent, naturally determined growth factors, and a raft of shifting environmental influences. What's more, these stages of development are at best statistical averages as opposed to ironclad absolutes. There are many variables in this development, which means each of us is, in fact, the exception to the rule.

Figure 7.2 shows that climbing these rungs of the ladder up to full operation of the prefrontal lobes takes many years. In fact, the higher up we go on this ladder, the more complex the neural structure and the slower its development. In this ascent, the unfolding capacity depends partly on whether or not we receive the appropriate nurturing environment and model. At each new rung we open to a new repertoire of abilities. In the unfolding of these new abilities, we overcome the limitations and constraints of the previous rung, transcending them to go beyond. An infant is radically helpless at birth, then crawls about on the floor, and finally takes to its hind legs, charging about.

Of course, there are no guarantees built in to this promising sequence. Each stage may or may not bear fruit depending on any number of variables and chance possibilities. If we have the appropriate nurturing and supportive models as infant, child, adolescent, and young adult, we will move into ever higher stages of conscious awareness and possibility as a matter of course, experiencing each new unfolding as discovery of a whole new world of delight and possibility—a time of breathless excitement.

THE SEQUENCES OF MATRICES

We can also view this developmental ladder as a series of matrices. In our first matrix, or source, the mother's womb, the so-called trimesters roughly correspond to the cellular growth of the triune or three-part brain. The first trimester gives rise to the reptilian sensory-motor system, with its initial stirrings serving as the quickening noticed early on by the mother. The old and new mammalian brains then arise in the second and third trimesters, respectively.

A major discovery of late-twentieth-century science was that the cellular growth and potential capacity of this triune system, as well as all subsequent developments, is critically dependent on the mother's positive emotional state. The "molecules of emotion," as neuroscientist Candace Pert calls this state, have a profound impact on the DNA involved in cellular structures right from the beginning of a fetus's development.

The second major matrix is the mother's arms. Nature intends that the infant shift from inside to outside that primary matrix, the womb, to allow for the growth of that new human brain, the prefrontal cortex, which grows throughout the nine months following birth. At the same time that the prefrontals begin their growth, the earliest reptilian brain begins its active development. The newest brain is able to enter as the oldest gets to work in large part because a mother's arms are the ideal matrix for the materials and models both systems need.

The prefrontals grant us humanity beyond all animal forms, but their growth is largely determined by the mother's emotional state during this in-arms period—and by whether or not the infant is in arms. It is important, then, to take care of a pregnant mother, to allow a mother to birth her infant with care, to take care of a nursing mother, and to take care of both mother and infant thereafter. Any form of emotional disturbance in either the lifegiver herself or the new life she gives can throw a monkey wrench into nature's plan. Those "molecules of emotion" are critical ingredients for growth when positive and can pack a destructive wallop when negative.

A child's success in the many matrices he enters after mother's arms determines the success of evolution's intent for the prefrontal cortex, designed to develop into the governor of the whole affair, keeping in line those lower animal structures that support the higher structures for all that nature holds in store. Being the latest of evolutionary systems, the prefrontals are the most fragile: Interference or accident anywhere along the line of development compromises their development. The importance of Maria Montessori's observation that "a humankind abandoned in its earliest formative stage becomes its own greatest threat to survival" is borne out around us every day.

It takes many years and many rungs of the ladder for the prefrontals to develop from the in-arms period to their full operation. An individual spends this time embedding and emerging. We embed as a psychic structure, personality, or mind, in each stage's matrix, complete the growth inherent to that stage, and emerge from it to embed in the next.

The third matrix is the family and immediate environment, the time of the toddler, which centers on development of the old mammalian emotional-cognitive brain. Again, the emotional atmosphere is the most critical of all elements to successful embedding at this stage. Here our abilities for memory, learning, immunities, and a whole list of capacities unfold to give us the ability to relate to self and world.

We emerge from the third matrix into the fourth, centered in the right hemisphere of the new mammalian brain and its semidormant prefrontal cortex, opening to awareness of the whole of nature-out-there. This is the time of Jean Piaget's "child of the dream" or Rudolf Steiner's "ethereal child" developing divine imagination. Through this fourth matrix a child from the late third to the sixth year can modulate his sensory environment, given by those two lower brains, and create a play world over which he has dominion. Through his play he develops the imaginative capacity on which all subsequent intelligence is critically dependent.

Emergence from this ethereal, dreamlike, right-hemisphere world leads to the fifth matrix, resulting in embedding in the left hemisphere of the brain. Between ages six and seven, the child experiences a huge brain growth spurt that leaves him with a neural mass in the skull that is five to seven times what it was before this point. A fundamental shift of awareness follows this neural growth as the matrix shifts to the left hemisphere of the brain. With this shift a new realm of thinking is opened in which the mind can operate on the materials of its physical world, selectively change aspects of physical matter itself, even reverse the ontological constructs of our living world according to what can be imaged in the mind's eye. Whereas before the child projected his inner image on an outer image and played in the modulated world of his own making—one foot in the inner world, one foot in the "real" world—in this new concrete operational stage, as Piaget termed it, using the same imaginative inner image, he has both feet in the ordinary world. Now he can modulate the nature of the outer object itself: he can make things, make music, invent, weave, garden, bake cookies—in short, again and again transform one thing into another.

From this concrete operational stage the child emerges into the sixth matrix, where the mind embeds in a process rather than in a place or thing. Piaget called this process formal operational thinking. This is the first stage of emergence of mind from brain itself, where we move into a realm of thought that can create systems beyond all physical boundaries. As we embed in this stage, we can reach back from our evolutionary roots to change the very brain structure giving rise to us, eventually discovering the way a strange loop forms. We can move from the concrete to the abstract and play with the boundaries between the two.

Formal operations lead to midadolescence and the mind embedding in the prefrontal cortex. To prepare for this evolutionary leap, the prefrontal brain undergoes a growth spurt that begins around age fifteen and isn't complete until around age twenty-one. this is an interim period of process formation and exploration, a loose-ended, not-yet-stable state called adolescence. At this stage, emotions are all: the adolescent brain-mind system calls for as much patience, tolerance, and nurturing from family, society, and world as the toddler did years earlier, and for strikingly similar reasons.

ON BIRTH AND BONDING

Shortly before I was scheduled to speak at a conference on birth bonding, I received a lengthy form from the American College of Obstetricians stating that I had to disclaim via my signature any conflict of interest I, as a presenter at the conference, might have with the practices of the College of Obstetricians. (It seems this group was cosponsor of the conference.) On the form, I was warned that my failure to comply would result in the College taking various steps, including informing the attending audience of my noncompliance, which would supposedly undermine my credibility.

The American College of Obstetricians, consisting of some forty thousand OBs, had just passed a resolution that any woman desiring a Caesarian section could have one, with no medical reason necessary. Interestingly, both hospitals and obstetricians make far more money on C-sections than on ordinary vaginal deliveries, and after heart surgeons, OBs are the highest paid physicians, with birth accounting for over half of all hospital revenue. Further, it is a fact that the majority of hospitals are owned by large corporate chains and the American Medical Association is a political power to be reckoned with. (Needless to say, I disregarded the offensive threatening form and have since spoken out at conferences and workshops as usual on the wide-scale damage wrought in hospital delivery.)

A Conflict between Nature's Intelligence and Human Intellect

Paul MacLean has outlined a triad of needs that are imperative to the well-being of infants and children and their development: audio-visual communication, nurturing, and play. All three are established by mother-infant bonding at birth and are stabilized through breastfeeding in the first year of life. Deprived of bonding, primarily, and breastfeeding, secondarily, all subsequent development of both infant and mother is compromised.

Years ago, child psychologist Muriel Beadle asked why it is that the human infant seems born in a state of alert excitement that quickly reverts to distress followed by conscious withdrawal, which lasts for ten to twelve weeks on average before full consciousness manifests in the infant. I suggest that an answer to this question is found in the hospital delivery of infants.

Most female mammals, on preparing to give birth, seek out the most hidden, preferably dark, quiet, and safe haven available. At the first sign of any intrusion (in the wild even the snapping of a twig brings about a startle-alert response), the creature's natural intelligence slows or even stops birthing as she waits to make sure the setting is safe. Mammalian instincts are in charge of human birthing as well, interpreting and responding to environmental signals much as animals do. At the outset, the hospital environment itself slows the birthing process dramatically in even the best scenarios. If a mother-to-be is safe, supported, and secure and is in touch with herself and nature, she can give birth in as little as twenty minutes. But at the first sign of any interference of any sort, regardless of the nature of or reason for it, the birthing process is disrupted, slowed, or even halted, by an internal ancient and powerful intelligence.

If disruption does occur, the mother's smooth muscular coordination of resonant responses can be lost, resulting in a kind of internal chaos—muscle fights muscle, instinct battles instinct, all with the attendant predicted and long-expected pain. Inner knowing is confused by outer intrusions, nature's intentions clash with culture's attentions, and mother and infant lose on all fronts. Sadly, this has been the "normal" birth process for the majority of women in our culture and a primary cause of our ever increasing personal and social turmoil.

Time in Arms

Nikos Tinbergen, Nobel laureate in ethology, studied the metabolism of the early infant and determined that a human newborn needs to feed about every twenty minutes in its early days, with the periods between feedings growing progressively longer as the months go by. Human mother's milk, it seems, has fewer fats and proteins than that of most mammals and, though rich in hormones, requires that the infant feed frequently—which is the whole point. Hydrochloric acid, necessary for the digestion of fats and proteins, is abundant in other mammal infants but is less prevalent in human infants because, it has been suggested, it is less needed (though this reason has been the subject of recent debate). Some mammals, rabbits for instance, produce milk so heavy with fats and proteins that their offspring need to feed only once or twice a day, allowing mothers time to forage freely and make more rich milk for the next feeding. Nature arranged the human's feeding schedule to balance an intricately woven fabric of interdependent needs the satisfaction of which is critical to being fully human. These center primarily on the emotional systems of both infants and mothers. Pediatrician Maria Montessori noted more than fifty years ago that at about nine months after birth, full-strength hydrochloric acid appears in the infant's system—and this nine-month marker plays a significant role in bonding. Steiner likewise emphasized the critical role of breastfeeding in the first nine months after birth.

Just as it took nature nine months to grow the infant in the mother's womb, it takes another nine months in the arms of the mother to establish that infant firmly in the matrix of its new world. Most important, as we have seen, at this time the prefrontal cortex grows. Further, nature must activate and stabilize all body functions, particularly the heart, which requires both constant reciprocal interaction with the DNA complex of the mother's body, heart, and emotional system and overlap of and entrainment between the mother's and infant's electromagnetic fields. To accomplish all this, nature does what she can to keep the infant primarily in arms for that period of critical brain growth and system stabilization.

In MacLean's triad of needs, the first is audio-visual communication. Rudimentary hearing develops early in utero. If the fetus has normal hearing and a speaking mother, language development gets underway in the second trimester, through muscular responses the infant makes to the mother's spoken phonemes, those foundational units of words on which all language is based. This phonetic-muscular foundation builds in successive stages until birth and quickly leads to speech.

Vision, however, which occupies more of our brain than all other senses put together, obviously can't develop in utero, though visual sensitivity appears early on, as observed in a fetus's aversion to bright lights shone directly on a mother's belly (which prompts the fetus to turn its head away). Full visual development, however, and the audio-visual communication that accompanies it, must await birth to unfold. At birth, if a face comes within six to twelve inches away from an infant's face, the newborn exhibits two immediate responses: its initial excited alertness stabilizes and reciprocal audiovisual responses between infant and mother begin, with full consciousness quickly following. The infant is born with a preset neural pattern for perceiving or cognizing a face and will lock eyes on a face immediately at birth, if presented with one at the requisite distance. The newborn will then hold that face in focus, which literally turns on the consciousness of that infant brain and keeps it turned on. Perception and cognition automatically begin, activating the infant's entire body-brain system. Parallax (muscle coordination of the eyes) forms within minutes (so the infant can even follow that face if it moves around) and the infant begins constructing knowledge of a visual world—all based on this stable foundation of a mother's face.

Before long, the infant registers other objects in the mother's immediate vicinity, and, through processes of association, corresponding new neural patterns of perception form. As a result, a cognitive field of recognizable objects grows exponentially (as does the brain itself)—so long as that face-pattern remains the stable point of reference.

Although any face will work at birth, face constancy and all that goes with it is the critical factor in an infant's early movement from known to unknown and is vitally necessary for a stable and stress-free development. Should a face not be presented at birth, distress takes over within about forty-five minutes and conscious awareness fades. Ordinarily, full consciousness does not reappear for upwards of ten to twelve weeks on average. Bonding as a reciprocal function between mother and infant is then fragmented and the ongoing nurturing instincts that bonding awakens and locks into the mother's responses also fail to open and develop. Most infants then receive only sporadic exposure to a face or faces. By this time, with the infant's consciousness largely retreated, it is missing the awareness needed for such face cognition to take place and become stabilized. Nature compensates as best it can, but under these conditions, its capacity to compensate is diminished and slow.

Nature arranged that this magical face trigger is also some six to twelve inches from the breasts, from which flow nurturing nourishment. Frequent nursing, then, assures both a frequent reinforcing of the stable face pattern upon which an infant's vision and awareness are based and continual stabilization of infant heart frequencies by entrainment with those of the mother. Object constancy, as Piaget called it, the stabilization of an object-world of vision, occurs around the ninth month of this busy construction period. Likewise around this nine-month milestone, if bonding and nurturing have been given the infant, myelination of the neural patterns of its primary visual world takes place, an object constancy that makes permanent the neural foundations of vision so that ongoing expansion of the visual world becomes automatic and effortless. Now nature can turn her building energy to other developments that begin around the pivotal ninth month.

Interestingly, any society that separates mothers from infants at birth has a disproportionately large population with impaired vision. The United States, for instance, is virtually a nation of eyeglass wearers. We forget (or ignore) that "primitive" peoples have far more accurate and extensive vision than we have—indeed, some of these people can see the rings of Saturn with their naked eye! Far more significant is the large number of infants and toddlers who, pushed about in various wheeled devices that keep them separate from their caregivers, have strangely vacant, barely focused eyes and vapid, nobody-at-home expressions, as though a light was never turned on in their brain.

Heart Bonding

About forty years ago, research at the University of Adelaide pointed out the now well established notion that the mother's heart is a most critical factor in a fetus's development from conception through birth. But the same close proximity of the mother's heart is every bit as critical during the first in-arms nine months of an infant's development. Over a half-century ago, researchers found that a heart cell removed from a live rodent's heart and placed in an appropriate nutrient to keep it alive continued to pulsate, expanding and contracting regularly according to the rhythm set by the donor heart. After being separated from the heart for some time, however, the rhythmic pulsation deteriorates until that erratic, jerky spasm called fibrillation, precursor to the cell's death, sets in. When two heart cells are placed well separated on a slide and fibrillation begins, bringing the two cells into close proximity (they do not have to touch and can even be separated by a tiny barrier) stops the fibrillation in both and reestablishes their coordinated pulsation. It seems, then, that each cell lifts the other out of that fibrillation that leads to death and into the shared rhythm of life. This is the core of the bonding phenomenon nature arranges at birth.

This cellular communication miracle occurs when the electromagnetic fields that arise from and surround each heart cell come into contact with each other, wherein their waves entrain, or enter into the same coherent pattern, lifting those cells out of chaos into order. Cells and their electromagnetic (EM) fields mutually give rise to or influence each other, and the same phenomenon occurs, on a far larger and far more serious level, with infant-mother hearts at birth.

As we have learned, the heart itself produces a powerful EM field in three successive waves: the first and most powerful surrounds a person's body, flooding every cell and neuron of that body and brain; the second extends out some three feet in all directions and interacts with other heart fields within that proximity, a principle ingredient of emotion and interpersonal relationships; the third extends out indefinitely, universally, which is possibly an aspect of the human spirit.

Following the separation of birth, infant's and mother's hearts must be brought into immediate proximity again in order to lift each other into their familiar, stabilized order, the resonance the infant has imprinted on a cellular level from conception. The infamous postpartum depression is an actual and quite real aching of the mother's heart itself. Again, that six- to twelve-inch distance of the mother's face, which also gives the infant immediate proximity to the nurturing breasts, assures the ongoing stabilization of the infant's and mother's hearts. By the ninth month after birth, the infant heart has matured enough to stand on its own, so to speak, without frequent restabilization.

Newborns and mothers wired for heart- and brain-wave recordings (via EKGs and EEGs) show coherency and entrainment (matching of the wave frequencies) when infant and mother are together. If prolonged separation takes place, both systems become incoherent (chaotic), causing general stress and the release of cortisol by both the mother's and child's systems. Excess cortisol is quite toxic to neural systems, particularly new ones; thus any society interfering with natural bonding at birth will have a corresponding increase of heart trouble in their adult population.

The Necessity of Touch

Years ago, Ashley Montague wrote a now-classic work called *Touching,* and recently writer and masseuse Mariana Caplan wrote *Untouched,* both well-documented studies showing the critical necessity of infant skin stimulus at birth, when the newborn's peripheral nervous system is quite undeveloped. The millions of sensory nerve endings distributed over a newborn's body could not be activated or developed in utero—in that water world the *vernix caseous* is a waterproof, fatty coating that protects the fetus's body and also insulates its myriad nerve endings. At birth, then, all animal mothers vigorously lick their infants off and on for many hours, even sporadically for days, to activate the newborn's dormant sensory nerve endings, for this peripheral nervous system is a primary extension of the brain and brain development is at stake.

Failure to activate these nerve endings results in a desensitization affecting, among many things, the reticular activating system of the old reptilian, sensory-motor brain, where all sensory stimuli is organized into those resonant patterns that are then sent to higher cortical areas of the brain for world-making and experiencing. Touch deprivation results in compromised overall neural growth and a diminished sensory system and general conscious awareness and affects inner ear development, balance, spatial patterning, and even vision later on.

Mothers separated from their infants at birth obviously can't provide this touch stimulus, nor are they themselves called to do so later if the separation is prolonged, for the mother, too, has a critical window of opportunity for activation of nature's ancient nurturing responses, our "species survival instincts," as Paul MacLean terms them. These touch instincts are activated by the mother's skin-to-skin contact with her infant, another facet of the reciprocal relationship of bonding.

Bonding Produces Language

Language learning begins late in the second trimester of pregnancy and is manifested as muscular responses the fetus makes to the phonetic content of the mother's speech. If the appropriate model-stimulus is provided—a speaking mother in close proximity—this dynamic continues after birth. During an infant's first nine months of life, with continued language learning and phonetic completion and mirror neurons busily at work, speech preparation takes place. At around the ninth month after birth, an infant's open imprinting of phonemes closes to those of the mother's language and the average infant's speech preparations result in infant babbling and even the first words. All of this occurs automatically if an infant is nursed, nurtured—and spoken to.

Infants separated from their mothers at birth and confined to various forms of ongoing separation thereafter (as most modern infants are through cribs, bassinets, carriages, playpens, strollers, and the lengthy separation of daycare) are denied all these responses, and their development is correspondingly compromised. Nature will compensate as best she can—but compensation is always a poor substitute for spontaneous, mimetic growth.

A New Brain

Perhaps the most important development of all these nine-month markers is the completion of the primary phase of the prefrontal cortex. Only this newest system can organize the entire brain into a smoothly synchronous intention, linking all our lower instincts, as well as thinking and feeling, with higher fields of intelligence and translate all the higher human attributes such as love, empathy, care, and creativity into daily action. The prefrontal cortex gives us what Elkhonon Goldberg calls "civilized mind"—but it develops only in a climate of love and care.

Allen Schore's research shows that the genetic structure of the prefrontal cortex proves to be the most experience-dependent of all brain systems—that is, the most critically dependent on appropriate environmental feedback, which is given through the multileveled functions of infant-mother bonding and relations and the overall positive emotional environment that results. Nurturing through breastfeeding, sufficient movement and sensory stimuli, immediate proximity to the mother's face and heart, the continual coherent resonance between mother and infant heart fields, language and speech stimuli: all these play an essential role.

Failure to provide this overall emotional support inevitably results in a compromised prefrontal cortex, which literally cannot grow sufficient cellular structures and make the necessary neural connections with the rest of the brain for full operation. A compromised prefrontal cortex results in an impaired emotional intelligence with a corresponding difficulty in relating with others or controlling our ancient survival reflexes. This failure lends to a corresponding tendency toward apathy, hopelessness, despair, and the many forms of violence we see and know around us.

The Orbito-Frontal Loop

All the strands of nurturing outlined here gather their forces to completion around the milestone of the ninth month of life. Then, from the ninth to twelfth month, another major neural structure grows to connect the new prefrontal cortex or executive brain with the ancient limbic or emotional brain and defensive reptilian brain it will eventually control. At the same time, direct, unmediated neural connections to the heart are intricately involved in this network. This major bridge between old and new brains, this orbito-frontal loop, as it is called (because it is located behind the orbits of our eyes), proves the most decisive factor of life and is critically experience-dependent. If emotional nurturing is lacking or fails, this bridge is compromised and our ability to attend, focus, control emotions, and relate to others will be impaired.

At the same time that the orbito-frontal loop begins its massive growth, the ancient cerebellum in the back of the brain undergoes a corresponding growth spurt. Rudimentary until this time, this cerebellum is involved in all speech, walking, coordination of muscular systems, and much more.

So, after the crucial period of birth and bonding, with its related developments, the infant arrives at the toddler period, which ushers in a new series of bonding with new matrices—as mentioned above, those of the family, society, the world itself—all of which lead to the continuing of the cycle of partnering, birthing, bonding with new offspring, and on it goes.

Marshall Klaus spoke of an interlocking "cascade of redundant patterns" provided by nature to assure this critical first bonding and called bonding the establishment of the greatest love affair in the universe, one on which this wondrous unfolding of human life depends.

Now we can see the astonishing and thorough intelligence and careful planning of myriad critically timed and interdependent responses that nature evolved over eons of time and invested in this birth and bonding process. And now we can see the astonishing extent to which modern practices have bypassed, compromised, or even eliminated this incredible architectural design of evolution. We can understand why our medical interference with birth—taken as axiomatic and unconsciously accepted as the norm—is lending to our global undoing.

The ruinously expensive takeover of all birthing by hospital and medical procedures has brought into play at the same time an equally expensive therapeutic operation to repair the damage we are causing. We can witness the strange contradiction of a nation caught up in patchwork healing and hoped-for wholeness while it allows a radically damaging, unnatural birth practice and its complement of daycare to continue unquestioned and unchecked. There may never have been a "golden age" of birthing and child-rearing, but there are also no historical precedents for a species abandoning its own offspring, as occurs today worldwide.

But there lies a path in bonding throughout the embedding and emerging process in all matrices as the prefrontal cortex fully matures. Our emerging from and separation from all physical matrices as we grow takes place, not only through the prefrontal cortex, but also through the heart, and is designed to be a lifelong exploration of the higher worlds. Eventually, this could give the means for moving into the universal matrix of creativity itself, beyond all boundaries and constraints of our physical realm, a handy shift to have available when the physical wears out its parts and we need somewhere to go.

8

BONDING: NATURE'S IMPERATIVE

As mentioned in the latter part of chapter 7, in his late years, neuroscientist Paul MacLean gave us a brilliant paper on the triad of needs that must be met at all stages of life—particularly the periods of infancy and childhood. This triad is critical to development and, as we have seen, can be summed up as:

- Audiovisual communication: This is not merely stimulus, but communication, a reciprocal face-to-face, heart-to-heart affair.
- Nurturing: In infancy, according to nature's imperative, this always accompanies the audio-visual encounter through breastfeeding (an important action far beyond that of simply giving nourishment).
- Play: This is as vital as any survival instinct and is an automatic response when the first two imperatives, audio-visual communication and nurturing, are met.

All three needs involve close, intimate relationship. We are born to relate with, communicate with, nurture, and play with our world and each other.

RELATIONSHIP AND THE HEART

Our heart maintains relationship between and within our cells, organs, brain, and body and the world of others. Appropriate relationship is intelligent action, and within the heart lies a neural complex or "brain" that plays a principle role in this ability to relate to our environment and each other in ways that lend to well-being and pleasure. This body, brain-mind, and heart dynamic determines the kind of world experience we have, which in turn feeds back into that dynamic. The heart is the command center in this play of relationship and mind is the reciprocating recipient.

Years ago my meditation teachers in India explained that while there are billions of egos occupying brain-bodies, there is only one heart. (I consider the word *ego,* from the Latin meaning "I" or "me," as sacrosanct here, not in the negative sense of the word as appropriated by culture.) The heart pulsing within me is the same as the one in you, yet each heart, without losing its universal nature, takes on the colorations or flavor of each ego-self sheltering and serving it.

The medieval Spanish poet Ibn Arabi, wrote, "[T]he heart can take on any form, a cloister for monks, or a meadow for gazelles." As an aside, the gazelle was a popular metaphor for the alluring female in thirteenth- and fourteenth-century Sufi poetry, and, while Arabi was himself a not-too-cloistered monk, he was famous for his spiritual and erotic love poetry. In nature's design, spirit and erotic experience are equivalents, inextricably intertwined, thus Arabi's heart was at home in either form without judgment, as it is equally within each of us.

The heart is neurally connected with every facet of the body and brain but has no neural complex for making judgments. The ability to qualify experience and judge the value of an event is assigned to the heart's servant, the brain in our head. As a result, the overarching intelligence of the heart rains equally on the just and unjust, leaving out of its domain our tangled fabric of judgment. To begin with, the heart doesn't register such information as content; apparently, it registers only an event's emotional nature. (We will find this same factor of indifference in the enigma of the Ground of Being or Vastness, which we will tangle with later.) Wisdom results when the judgmental capacity of brain is in synchrony with the heart's intelligence, resulting in discrimination, discretion, and logical common sense as passed down through generations. Judgment, on the other hand, is wisdom run amok, the brain functioning without the heart, leading to culture and its constraints.

Universal heart and individual mind are mutually dependent, each necessary to the other. Judgment of a person disrupts relationship, an unintelligent error the heart is apparently incapable of making, but which the brain-mind seems fatally subject to making unless it is bonded with that heart. Out of the reciprocal mirroring of heart and brain our universe arises, with our individual self as its center.

For each of us, our universal heart is that deep inner presence of *Me* that leads to a puzzle. The ego-self is the only way by which the heart can experience its world and see itself anew, moment by moment, thus no actual separation of heart and its ego-self is possible (by sheer biological default). Yet each of us often feels isolated and alone, despite whatever intimate relation with our heart we might have and despite our awareness and nurturing of this relation through meditation, HeartMath practice, yoga, silent prayer, and so forth. Often feeling incomplete, isolated, lacking, we seem driven to search for relationship out there in the world. Relationship is the prime driving force behind all creation, and our seeking it, regardless of inner orientation, is just as nature planned.

A common misunderstanding among spiritually inclined people is that if we were truly established in the heart, we would be completely self-sufficient, emotionally rich and full, dependent on nothing out there. Part of this misunderstanding is the notion that as long as we are at all dependent on anything out there and other to us, we have fallen short in spiritual development. If this was true, we would at best experience various forms of compensation that would leave our heart restless and unfulfilled, though our ego might be filled with pride of spiritual attainment. Relationship is all there is and what the heart longs for.

UNITY AND COMMUNION MUST OCCUR THROUGH THE HEART

The dynamic of the heart-brain-mind relationship in a dialogue of balance is the foundation, not the goal, of the great journey of relationship. Making a goal of our foundation—like assuming the rocket's launching pad is the fulfillment of the rocket's purpose—stops us at the foundation point.

The goal of full heart-brain-mind dialogue is simply reestablishing our basic natural integration before enculturation split our system. This does not imply a kind of Luddite, back-to-the-jungle time, but it may be that only by getting us back to some hypothetical point zero of our natural state can a movement into full relationship and development take place. Our natural state is our launching pad, not a target for our space probes.

Living fully is a journey into relationship, and sexuality is a most effective and reward-laden way. Our sexual nature assures this move out into the world, no matter how successful our inner world or self-love might be. No matter how bonded we are to our own heart or how sufficient within ourselves we might think we are, relationship with others out there in our world is critical to our heart's unfolding. Sexuality can bring a synthesis of all relational needs, pleasures, and joys of both mind and heart and can be a key to that kingdom within.

The fulfillment of the heart's relationship occurs through one heart relating with another heart. Because there is only one heart, however, it can relate to itself only as itself in another person. It is through this that the proposal that heart as both universal and individual takes on cosmic significance. In taking on the colorations and characteristics of that self or person in whom it is lodged, the heart has simultaneously both unique individual expression and universal ground. Two hearts, each experienced as an individual self in its unique variation, can experience ecstatic relations with each other through their individual selves, fusing into the universal aspect within each. Two hearts beating as one in their physical overlapping necessarily include a corresponding overlapping of their individual electromagnetic fields radiating out at the same time that the universal within is fusing. It is telling to note that we are the only mammals who make love face-to-face, thus heart-to-heart.

Also telling is Wilder Penfield's discovery that an individual, a personal mind-self in full, ordinary consciousness, can experience simultaneous bilocality. Biologist Gregory Bateson spoke of mind and nature as a necessary unity that is but another aspect of the same unity found in individual and universal, God and human, spirit and flesh, heart and mind, self and not-self. To function, the individual heart and mind must be separate and unique on one level, but on another level their unity, the ground from which their uniqueness springs, never changes. Only through their uniqueness can relationship take place, and only through relationship can reality be created and experienced. On this rests the whole riddle of existence, though it leaves hanging the primal paradox around which our life circles: As Søren Kierkegaard said in his great prayer, "Even the fall of a sparrow moves Thee, but nothing changes Thee, O Thou Unchanging."

Unity and communion can be found out there in others through the heart. Heart longs to bring each *Me* into relationship with another *Me*. If we are consciously connected to our heart and aware of this communion within, we find this community out there wherever we are, and any individual relationship, sexual or not, is then rich and full, a representation of the whole.

If our heart-mind connection is compromised, however, this reflects outwardly as well. We then see other individuals through a prism of that compulsive cultural priority demanding "What's in it for me?" or "How can I use this situation or event to my advantage or for my welfare or self-protection?" We unwittingly prejudge events. Searching for our gain or advantage, we see nothing else. What our present moment offers is not cognized. When we rule out the heart, communion is compromised and both mind and heart are the losers. If we experience enough of this, we will probably die of a "broken heart" in any of its myriad guises.

From conception, our enculturation separates us from that heart-self within (which Robert Sardello calls the "soul of the world") and prevents us from, through our head, getting back in sync with that universal within us. Which of us can, through thinking, increase our stature by a cubit? There are many mind-manipulation tricks for sale, but, however tired the truism seems, only opening to the heart can do the trick. Sardello's soul practice can prepare the grounds of mind for the letting go necessary, as can applying HeartMath practice. But like the final step in Laski's Eureka! process, the mind must get out of the way for heart to reestablish the broken link to real joy in relationships, the real purpose of life and the driving force behind creation and the universe.

The claim that pleasure and joy are the whole purpose of life is hotly contested by our entire cultural ideation, our very mind-set, and, above all, religious belief. We are here to serve culture under its myriad guises: worshiping god, serving humankind through self-sacrifice, suffering and working for the welfare of those out there, and so on. This world of cultural counterfeits clamors for our attention and drains our lives. Sooner or later, we find each counterfeit an empty delusion, a virtual reality that leaves us with nothing but our unassuaged longing heart. If, however, we were truly bonded with our heart, any of those counterfeits would reveal the truth behind them and we would truly serve the heart and bring real relationship by investing our life in them. Intent outweighs all other factors, and the intent of the mind united with the heart unifies all culture's fragmentations. As Kierkegaard said, purity of heart is to will one thing.

THE ROLE OF DNA IN MIRRORING

Because relationship is the only way anything can be, nature has a biological imperative for our development. To be initiated and developed, each infant-child's potential ability, even those genetically inherent in us, must be given direct contact with an example of that particular potential in its developed, functional form. Only if given resonant signal information from a model can the DNA within the infant-child build its own fair copy, because this building involves mirror neurons and resonance between the DNA of child and model. We are born mimics. Through imitation new life and creativity arises.

In this we find why, as mentioned by those investigating mirror neurons, virtual realities betray the developing child. Resonance between entities, Paul MacLean proposed, is the way all relationship and events unfold; the way the brain communicates its many relationships; indeed, the way the whole neural system of body-brain works. Resonance is relationship, and relationship, again, is all there is. If you pluck a string on an instrument, a faint humming occurs in all strings resonant with the one plucked; they vibrate through the waves set up by that primary string. The whole physical basis of harmonics rests on this relationship, and when we walk into a room and note that the "vibes" feel good we are simply experiencing resonance. Virtual realities involve frequencies that can stimulate certain audiovisual and positive-negative emotional reactions but offer no resonance, nothing synchronous with the heart's feeling-sense, nothing similar to the resonance we pick up in wordless communication with others, the feelings that can nurture, comfort, and strengthen us. Feeling and sensory stimulus are hardly identical.

Years ago, a research group had some rabbits that were being used to test various drugs and vaccines. These animals were inoculated with viruses to see which new pharmaceutical wonder might keep them alive the longest. The victims were kept in separate cages and all proceedings were carefully followed and noted in the records. One particular rabbit remained miraculously healthy throughout testing, fat and fit, its coat glowing regardless of the deadly viruses pumped into it—indeed, in its case every drug seemed a wonder drug. Then the mystery was finally resolved: The night caretaker of the animals had taken a fancy to this fortunate little creature, and rather than just feeding it, he took it out of the cage each evening and hugged, petted, and played with it. DNA is acutely sensitive to signals from its environment, particularly those "molecules of emotion." The emotional centers that control the immune system are extremely sensitive to these molecules broadcast from every direction.

A 1998 study shows a similar phenomenon: Briefly, it states that a pregnant mother's emotional state critically influences the brain growth of the infant in her womb. If the mother feels secure, loved, calm, and cared for, or—of paramount importance—if she can establish such a state within her regardless of environment, she will give birth to an infant with an enlarged forebrain (the source of higher intelligences) and a smaller hindbrain (the ancient reptilian survival system). If a mother is depressed, fearful, or anxiety-ridden, she will give birth to an infant with a heavier musculature, enlarged hindbrain, and reduced forebrain, a commonsense move of nature's, though a devolutionary one.

If DNA were not sensitive to information signals from its environment, evolution could never have taken place and our remarkable adaptability would not exist—and the higher in evolution we go, the more critical this DNA-emotional factor is. At every conception in our species, then, nature asks, in effect: "Will we be able to move into the higher realms of intelligence this time, or must we defend ourselves again?"—an evolutionary question answered by the molecules of emotion in our mother.

THE HEART AS BRAIN

The heart is a major endocrine gland of our body, producing a whole family of hormones that affect the functions of every organ and, above all, the brain. Which hormone is released when to influence what is determined largely by information sent to the heart from the emotional-cognitive or old mammalian brain in our head, as well as from our body as a whole. The heart has a complex neural structure or "brain" that connects with the brain in our head and every organ in our body. Thus the heart literally orchestrates the whole complex of body-brain-mind. (See figure 8.1.)

As we saw in chapter 5, the heart is also the source of a powerful electromagnetic field of energy encompassing us and saturating every cell in brain and body, just as the electromagnetic fields surrounding the earth saturate all things living on the planet (see figure 5.5 on page 76).

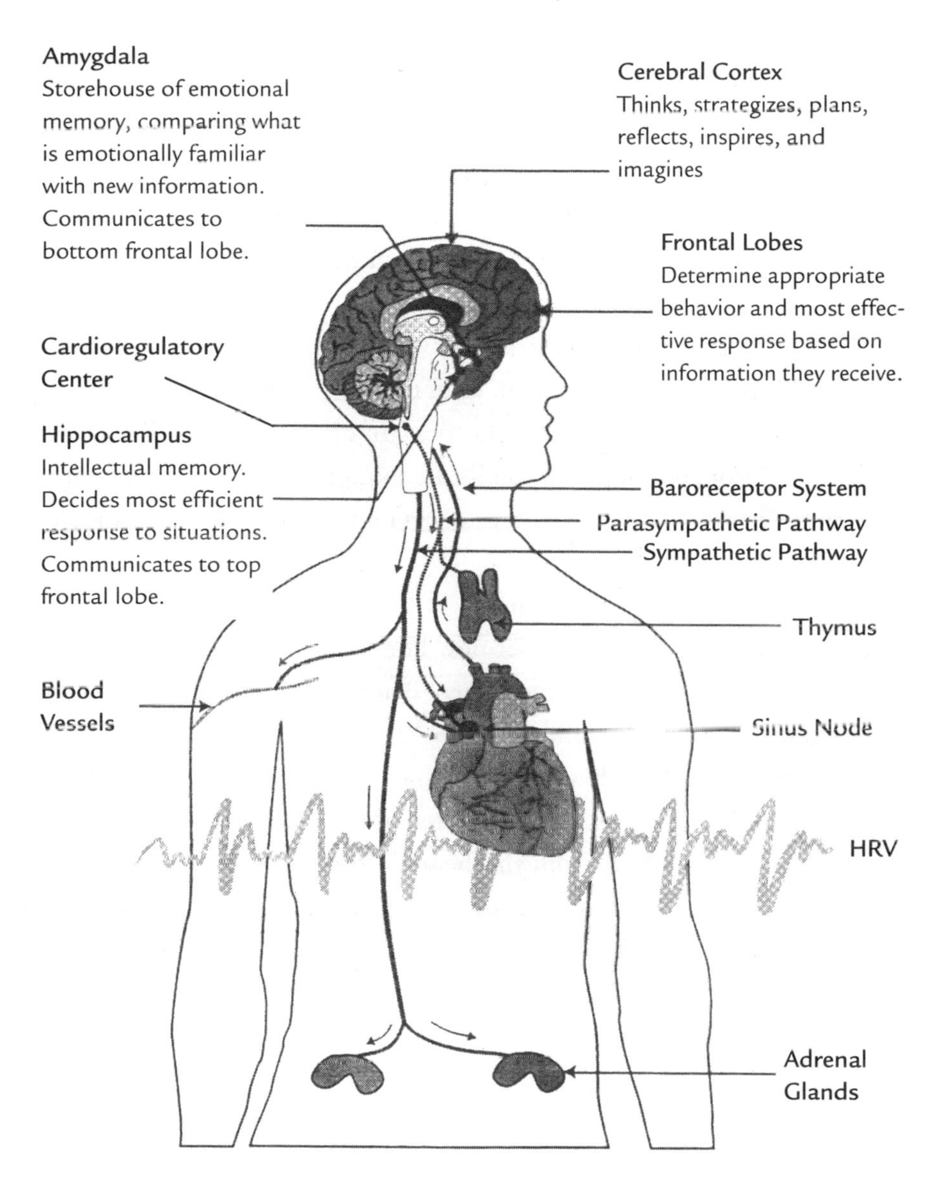

Figure 8.1. In this diagram of the heart-brain dynamic, note the location of the amygdala, which plays a principal role in emotions.

Robert Sardello asks whether this electromagnetic play is the same as our spiritual energy. The relationship between this electromagnetic field and cells is probably similar to that between mind and brain and is, at some point, a strange loop.

Rudolf Steiner predicted that the greatest discovery of late-twentieth century science would be that the heart is not merely a pump but also a major source of intelligence and that our greatest challenge would be to allow the heart to teach us a new way to think, which, it seems, would open us to higher worlds. Steiner pointed out that the heart picks up and responds to both the inner senses of our body and the outer senses of the world. In order to view the world anew, Sardello says, "the 'task from the heart' is [to approach] everything in the world with the sense of not-knowing, as if discovering it for the first time"—that is becoming as a little child. Dropping our ideation and assumptions thus can reconnect us with the heart, a first step to ever deeper and more conscious connections with our life.

In actuality, we wear our previous knowing like a shield, thereby insulating ourselves from that ever-new bonding and life the heart brings about and offers moment by moment. "Behold," Jesus, that great being of heart, said a couple of millennia ago, "I make all things new." But we don't want all things made new; we want to correct some aspects of our present moment—generally the behavior of others. In order to open to the newness of that flow from the heart into our present moment, we must leave behind the record of all the injustices done us, our wounds and scars of relationships, our angers and grief for which we demand redress or justice. Hanging on to these old events, we harbor our chaotic history of injustices as a sacred trust, though they tie us to that past and bar the new life we may long for. Further, if we want something other than just a better hand dealt from the same worn-out deck of cards, we must leave behind our notions of what we think that newness of mind must be or might be.

In our enculturated blindness, however, we screen out this present moment on behalf of a constant replay of past knowing, like a religion bending our present back onto a dead past or implanting a dead past into the present. We are, as Sardello says, continually forgetting that we experience a unified world. The heart remembers this unity—in fact, it puts its members back together instant by instant, even as they simultaneously separate to relate to each other, giving us our world. Steiner wrote of the "chakras of the heart" and our reawakening to them, while Sardello speaks of ways to awaken—and, I would add, to awaken to—the forces of heart. All is reciprocal, nothing unilateral.

THE BUILDING OF LIFE

A test-tube baby offers a good example of resonance between DNA and the models imperative to life. First, the term *test-tube baby* is a misnomer in every way, implying something impossible. An egg and sperm can be brought together in a test tube, and union may take place there, but after a few dozen preliminary divisions of that fertilized cell, it then dies unless placed in a womb (or frozen for future placement). DNA makes its initial gesture toward mitosis in this cultured situation, as DNA is programmed to do for all forms of cellular growth, but to build a new life requires a living mother matrix. Following that initial built-in division is a mitotic chain of incalculable complexity that goes on and on—up to seventy-eight trillion cells acting in synchrony, in fact—until a new life is created.

Scientists have tried to find the appropriate nourishment and temperature to keep that dividing cell going in that test tube, but to no avail. Think of it: If they could keep that embryonic magic going all the way to completion, making a genuine test-tube baby, then voilà!, those magicians would have successfully played God on the largest scale yet, beyond even the scale achieved with the making of the atomic bomb. And, while success in such a glorified sleight of hand might well create a nightmare equal to or greater than that created by the bomb, bear in mind that the successful manipulators would almost surely win a Nobel—an ultimate prize for which many a scientist would sell his or her soul or that of their mothers or of Mother Nature.

In order to build a whole new life, rather than just a handful of loose test-tube cells, the newly conceived DNA must receive audio, hormonal, electromagnetic, and many other frequencies from the mother's heart. In her wisdom, then, nature placed the DNA womb adjacent to that mother's heart (though it may have taken an untold time of Darwinian random-selection to work out the details).

In addition to this heart-DNA dialogue, every organ and part in the mother's body also sends the DNA its model information. This results not just in an infant's fair copy of the maternal heart but also of the liver, lungs, stomach, and so on. If no models are provided, as in that paltry test tube, none of this can occur. DNA can't create a fair copy of a test tube.

THE MODEL IMPERATIVE FOR LANGUAGE

Decades ago, Noam Chomsky claimed that language was far too complex to be so quickly learned as it is in infant-children and must therefore be an inherent capacity with which we are born. Remember: Any action repeated long enough becomes a habit, any habit repeated long enough can become an instinct carried genetically and functioning below awareness. Such is the history of language.

Some time between the first and second trimester, the uterine infant's rudimentary muscular system reacts to sounds in its environment, and water conducts sound fairly well. Loud sounds bring a crude startle-alert response from the fetus. At this point, each layer of muscle tissue comes with a muscle spindle, a tiny speck of nerve stuff connected to a network of such spindles, with all connected to the cerebellum in the back of the reptilian brain. It is the cerebellum that is eventually involved in, if not director of, all language-speech functions, as well as muscular systems, balance, walking, and so forth.

In the second trimester, the emotional-relational old mammalian brain makes its initial response and the fetus's musculature begins to selectively respond to the mother's voice. In the third trimester, the new mammalian brain, precursor to thought and speech, enters the scene of action, and that once-random muscular response selectively narrows yet again to precise muscles or muscle groups that respond to correspondingly precise individual phonemes given, of course, in the mother's speech. This muscular reaction involves pretty much the whole body and is the physiological basis for human language. (William Condon and Louis Sander first mapped this repertoire of muscle-phonic response in newborns back in the 1970s, and in France the physician Alfred Tomatis performed similar research.)

As we saw in chapter 7, at birth the infant visually—and just as selectively—responds to a face. Though it will respond at first to any pattern complex enough to bear some similarities to a face, response to such substitutes is short-lived. The newborn's visual brain has a built-in pattern for facial cognition, the only preset pattern there. If activated by an actual face, this pattern turns on not only the visual brain, but conscious awareness itself. If nature prevails, a face is accompanied by a parent who furnishes the host of other stimuli in that triad of needs that the infant system is geared to receive. This face is thus the fuse that ignites the controlled explosion of a new life into its new world. Surely, any of us who have not had the fortune to lock eyes with that new life as it emerges from its first matrix and visually searches for that needed face have missed one of the many opportunities for rebirth that nature offers adults.

As we have learned, throughout the period following birth, close proximity to the mother's face (a distance of six to twelve inches) is provided by breastfeeding. This also places the infant in immediate proximity to the mother's heart and all her body parts, which furnishes the DNA of the corresponding infant parts—most critically, its heart, with continual reinforcements of the original stimuli. In such close proximity, the infant's facial and neck musculature mimics, in minute movements, those of the mother's muscles when she speaks. Worldwide it is found that mothers are prompted by nature to talk to their infants in the voluble manner unique to the mother-infant relationship, and much responsive brain growth and action results. There are, in fact, some two hundred small facial and neck muscles involved in speech; these are activated and practiced by the infant's automatic response.

Nature assures that some 80 percent of the infant's waking time is spent staring at that face during those first weeks of life. In this fashion, the infant's mirroring of the model's muscles connects the two main branches of speech preparation, one of which has already been established through the phoneme-muscle play in the uterine period, and the other from the audiovisual actions of the in-arms period following birth. If an infant is not in arms frequently enough, this development is compromised and delayed.

This commingling of body language and speech expression prepares for the lalling period, that time of nonstop babble during which these phonetic-neural connections are practiced over and over, often resulting in an infant's excited squeals of joy. (We all love the sound of our own voice, after all!) This practice myelinates these neural fields, making them permanent. The infant is then ready for actual word play of its own, which rounds out the full dynamic of audiovisual communication and the corresponding nurturing and play that takes place with it almost from the moment of birth.

Every step of this imitative interplay depends on the mother model. If she is a deaf-mute carrying a child, for instance, and the womb is essentially silent, other than digestive rumbles and breathing noises, only the most rudimentary muscular response to sound will take place in her fetus; no selective response to phonemes will form because none are modeled. The infant will be born with no muscular response to language and will thus be speechless until nature is given the opportunity to compensate.

Should breastfeeding not take place, the speech muscle phase will not unfold as designed. Any later compensation will have to compete with other patterns nature has, over eons, evolved to fire at their respective after-birth times. Speech will be indefinitely delayed and ongoing development can be compromised.

In the event of such model failure, full consciousness is also delayed, for that face pattern activates not only the visual cortex, but the entire brain and conscious awareness. The occasional glimpses of a face presented at bottle-feeding time or diaper change, for instance, is not frequent enough exposure to give a stable flow of information the DNA needs for development and myelination of those primary patterns.

In utero, as nature planned, the bulk of energy can go almost wholly to language-muscular imprinting. After birth, during the in-arms period, it can go to speech mirroring and visual development. These foundational capacities will develop fully and automatically if met by appropriate models at the time of opening, a stipulation true for all developmental stages throughout childhood. Even those that open at maturity and carry us onto higher worlds must have their appropriate model and nurturing to unfold.

Cellular biologists Scott Williamson and Innes Pearse pointed out sixty years ago, after research of more than twenty years, that if given appropriate environmental response at the time of any ability's appearance (from egg on) full function results, free of charge. Otherwise, function is lost and nature must compensate. But compensation, they made clear, is bought at a price that sooner or later can break down. Though we compensate on ever greater levels, that whole repair venture eventually collapses and death of all functions takes over, stage by stage, which can drag out for years, thanks to modern medicine.

Failed function, compensation, and death, Williamson and Pearse's unholy trinity fills the vacuum left by the unmet needs of MacLean's triad. Nature's intelligent design, it seems can be broken by human's unintelligent moves.

BOND BREAKING

Following the scenario developed at the end of chapter 7, consider what happens to bonding if an infant is born in a traditional hospital setting, which was the norm during most of the twentieth century. The large array of technical interventions that are often imposed produce a semiconscious mother and infant drenched in bright light and cold temperatures. The umbilical cord is cut prematurely, causing the infant oxygen deprivation and panic. No face is immediately forthcoming, only masks, followed by an array of violent maneuvers involving needles to ward off demon germs, scrubbing to wash off dirty womb deposits (yuk!), followed by the ultimate insult of isolation in the form of bottle-feeding and nursery crib. All of this robs the infant not only of the needed face pattern and all that goes with it but also of any chance for compensation. And worse, as a result of this initial lack, the many subsequent models that are designed to follow the first in sequence will be given only haphazardly if at all.

The biological imperative of bonding with appropriate models at birth holds at each of the various matrices unfolding throughout childhood. Should a benevolent, kind, and welcoming birth occur and a benevolent, kind mother offer nurturing of every kind thereafter, a civil, benevolent new life has a chance to form. This civil benevolence will be sustained and developed, however, only if followed by the supports of civil, benevolent family, earth, and social matrices.

Decades of research show that though medical technological hospital childbirth is a major tragedy in history, it is nonetheless a major cultural imperative. Variations of such intervention have been going on for generations, even centuries, as Suzanne Arms exhaustively details in her classic book *Immaculate Deception.* Dominator males usurping the power of women and taking over from them the shaping of an infant's critical entry into the world and the imprints that follow have cast their long, dark shadow for a long dark time. Unfortunately, the growing percentage of women in the medical world, including obstetrics, has had little impact on the travesty of medical birth. Too often women must ape the graces of their models, teachers, mentors, advisors, graduate examiners, and the like, if they are to be granted final, credentialed blessings and entry into the closed brotherhood of medicine. Culture is not easily displaced.

At best, early imperative bonding is compromised—even prevented—by culture, resulting in isolation, abandonment, compromised development, and so on, and the world's average citizen today. In fact, preventing or breaking these bonds is as primary a need for culture to survive as the establishment of these bonds is for a peaceful society to survive. Culture, as a learned and acquired body of survival strategies, conditions us from birth, and by default we then automatically employ these strategies all our lives, primarily to survive culture itself while ironically sustaining its demonic force.

Of all the vast web of commands, directives, prohibitions, and inhibitions backed by corresponding threats of possible punishment, that of abandonment, the severest threat the human can experience, is locked into place at this cultural birth scene. This, however, brings about in the infant-child those predictable and controllable behaviors that our culture finds so necessary. Abandonment is, then, the foundation of enculturation, and through assuring the infant's cultural conformity from the beginning, this imprint guarantees the child's own pattern of parenting later. Having been shaped by this cultural imperative, we know no other way of being parents ourselves.

Although we may often be at odds with our culture and although our conscious intent for good is genuine and sincere, at best the infant-child receives mixed signals from every direction. Such cognitive dissonance early on brings a split in that infant psyche, one that will widen with each passing developmental period. In this way, culture as a formative field grows in strength generation after generation, century after century, and the violence it produces expands accordingly. There is a way out, of course—not through the head, book learning, and behavioral modifications, but through the heart and its resonant fields of love and belonging.

9

THE BIOLOGY OF RELATIONSHIP

All living creatures, from the simplest cell on up, have some form of positive-negative sensory detector, essentially a chemical approach-avoidance response necessary to basic survival: Embrace (or devour) an event that nurtures; avoid an event that threatens before it devours you. The higher in evolution, the more extensive is this primary instinct.

In mammals nature evolved a distinct module of brain for the purposeful scanning of incoming information to determine its positive or negative character. Signals from all cells of our body send millions of molecular-chemical information bits each second, brought together at the top of our sensory-motor brain in a passageway called the reticular activating system, or RAS. Gateway to the higher forebrain of emotion, thinking, and conscious awareness, the RAS is, among many things, a kind of body clock with built in rhythms of sleep-wake. Open the gate, we wake up. Close it, we sleep. In case of emergency, the RAS alerts us through signals sent from a brain module called the amygdala. (See fig. 8.1 on page 129.)

THE AMYGDALA'S REPEATING RESPONSE

The amygdala lies at the critical area between the ancient RAS, high point of the reptilian sensory-motor brain, and the lower level of our relational-emotional old mammalian brain. The ancient amygdala (about the size and shape of an almond—thus its name, which means "almond" in Greek) is evolution's watchdog for higher intelligence. It is the central focus of a number of emotional sensors in that area and acts like an editor, screening that vast stream of information as it moves up the RAS to the higher brains for processing. Any information indicating something harmful, threatening, or unusual is somehow sensed by the amygdala which instantly sends an alert to both hindbrain and forebrain. Not only does the amygdala scan this massive input, looking for trouble and not missing a trick, but it also saves its work, selectively locking in negative impressions with which it will later compare incoming experience. Only resonance or emotional similarities, rather than specific content, is part of the amygdala's memory storehouse. Further, the extent, depth, and intensity of the amygdala's alert, sent both up and down, varies according to the strength or duration of the alerting information in the flow and the nature of the amygdala's matching resonance. Such alerts range from minor, pay-attention notices to all-out war mobilization.

Throughout the first three years of the child's life, the amygdala lays down its resonance repertoire of negative experience, and at around age three this selective storehouse myelinates, its function becoming permanent. For the rest of the child's life it will quietly compare current events with its negative template, informing the brain as needed. The vast sweep of this watchdog censorship and signaling goes on below our awareness. Mind, as usual, is the last to be informed. We are all familiar with frequent startle-alerts, from unexpected loud noises and the like, but much amygdala activity is subtle, functioning through automatic, habitual patterns of thought or behaviors of which we are hardly conscious until those responses are upset.

In chapter 10 we will consider a child's action that arouses a parent's swat on the behind and thunderous "No!" A parent might realize that such actions are the child's play as he investigates his wondrous world or imitates, as nature intended, some action of the parent. We parents seldom reflect, however; we only react. The amygdala is instinctual and instincts react but can't reflect. Upon reflection, a parent would find the child's seemingly provocative action to be resonant, no matter how far removed or different in actual content, with a similar pattern in that parent's own childhood, some negative event locked into their own amygdala that fires in resonant signaling. Again, content is not part of amygdala's repertoire (as it might be in memories from mirror-neuron's mimicking); it deals only in resonance between the emotional nature of event and information. Probing further, a parent might find that his or her negative childhood experience was brought on, in turn, by their parent's similar event in childhood, which was brought on by . . . and on and on it goes. Thus the father eats sour grapes and sets his son's teeth on edge, generation after generation, all beneath the threshold of conscious awareness, where only the fallout registers. Addressing or attempting to change child behavior may be futile, then, because the problem probably lies with the parent, with his or her mirror neurons and amygdala at work.

Studies show that by the time a negative alert reaches our conscious awareness, millions of neural responses throughout body-brain have already been set in motion. Our reflecting mind's only resource after the fact is through rationalizing and apologizing to ourselves or others for those automatic reactions of ours that "just happen." These lapses are more significant than we might think, however, because in such acts of rationalization, our higher brains are being incorporated into service of the lower survival brain, which is devolutionary. Centuries, perhaps millennia of such devolutionary patterns may have brought our species tottering to the brink time and again.

Of course, the amygdala records the resonance of events that really are threatening, hurtful, or damaging as well as those simple knee-jerks that occasionally provoke a twinge of conscience in a parent. Resonance has no logic for such niceties of evaluation; it has only its frequency-matching action with something similar. This similarity can be vague, half imaginary, perhaps illusory or delusional, but regardless, the reach of resonance is long. The amygdala's indelible memory casts its shadow over generations, and in this way, culture forms anew as a near-immortal legacy.

PAIN CENTER AND PLEASURE CENTER

In the amygdala are the roots of a most insidious and compulsive conviction we inherit and can't escape: that children must learn to behave, must learn to mind! On this compulsive directive culture is maintained and many a child's life is wrecked. If a child deviates in any significant way from our behavioral patterns locked in through our childhood experience, the amygdala-born reaction fires, bringing a near-instinctive and irrational determination on our part to modify that child's behavior in accordance with our own. Thus we imprint on our own children the same crisis-laden demand that was made on us, a negative looping of mirror neurons imprinting guilt and shame, with fear springing from the amygdala's resonance patterns.

Beneath all this is the notion that should our child not act accordingly or "behave," our acceptance and approval by our culture is itself threatened. If our child doesn't culturally conform, we are held responsible and, thus guilty, are censured by those out there. The unseen force that has from our beginning determined the general shape and action of our brain, with its mirror neurons and amygdala, plays its hand within us and spins the cycles of culture, generation after generation.

On the brighter side of sensing, years ago neuroscientist James Old discovered a reward or pleasure center in this old mammalian emotional brain. While this brain makes qualitative evaluations of hindbrain information, the reward center is a neural module that goes far beyond evaluations. On picking up frequency signals resonant with this module, it can respond with an endless variety of pleasurable experiences, ranging from minor titillations that delight us to rewards so intense as to be almost antisurvival.

In Old's experiments, for instance, rats implanted with an electrode in their pleasure center that they themselves could lever-activate would stop at nothing to work the lever and maintain the resulting pleasurable state. They would forego food until starved to death and water until they died from thirst, males ignored females in heat, and some rats even crossed an electrified grid to keep the electrode activated and that pleasure center going.

All mammals have this powerful impetus built into their emotional brain, and the more evolved and intelligent they are, the more powerful the center. As William Blake commented:

How do you know but every bird
That cuts the airy way
Is an immense world of delight,
Closed to your senses five?

Humans in whom Old managed to implant such an electrode (a difficult feat) reported that "all the bells of heaven rang" when that center was activated. Indeed, anything pleasurable we experience has activated this center to some degree. The stronger the pleasure or joy—as we humans find in sex and orgasm, for instance—the more powerfully the center is responding. Mystical-religious experiences may involve this center and drugs can activate it, making it as antisurvival for us as in the case of those poor rats in Old's experiment.

That our whole world and universe produces an endless variation of this rewarding frequency in an infinite number of ways and degrees is the only design this universal creative intelligence can bring about, because creation is, of its own, this joyful pleasure. That we have such a capacity built into us by evolution (as partially seen by Darwin 2) is far from random accident or chance; it is simply an expression of the raw material from which creation springs. What's more, relationship between products of this joyous process, such as you and me, reflects, magnifies, and enhances the pleasurable output at every turn, as when a relationship of ours is resonant with us and our center applauds and cries "More! More!" Surely the cosmos does as well.

That this default design of pleasure and joy underlies the very foundational purpose of creation, stochastically shadowed as it may be, is indicated everywhere for anyone with eyes to see. It underlies the personal conviction by which I write of it and gives rise to my general nose-thumbing to those tough-minded behaviorists and puritanical fundamentalists who have helped rob us of this, both our greatest legacy of life and life's investment in us.

All of this is to say we are born to enjoy this life in its every facet—but, like any other intelligence or capacity, our capacity for joy must be nurtured and developed. Nature's model imperative holds here as elsewhere. All that our hard-core scientists or fundamentalists have proved in their condescending but relentless dismissal of such soft-minded proposals as this is that such a wondrous gift of spirit was not nurtured in them. Culture has stamped out their spontaneous and natural joy. Why they then feel such compulsion to foist the paucity of their spiritual life on others, as fundamentalists in general tend to do, is strange indeed. Does a tone-deaf person feel compelled to eliminate all the great music in our world?

At any rate, what we interpret as pleasure and joy is the simple state of love within and by which the universe manifests itself. This great force is evolution's raison d'être, the stuff of that Darwin 2 love and benevolence bringing us about.

POSITIVE COHERENT WAVES

Every pleasurable experience we sensing creatures have activates this positive reward center within us to some extent, sending a coherent wave of energy that can sync, mesh, or blend with other coherent waves or fields. These positive coherent waves (incoherent waves are negative) are directly associated with the heart and are simultaneously reflected in the electromagnetic fields emanating from the heart—which is a significant part of the heart-brain interplay and is critical to every aspect of our life.

Each cell of our body apparently registers a form of pleasure-pain, as biologist Bruce Lipton makes clear. Thus, when experiencing pleasure or joy, our body as a whole radiates varying forms of coherent waves, a state of mind reflected immediately in the magnetic radiations from our heart. A spectral analysis of our heart field reveals one of these two major states reflected: coherent (positive) and incoherent (negative). Coherent heart frequencies arise from relationship, love, joy, and other such positive emotions and can, in effect, propagate or grow in strength or be reinforced by other coherent waves, which can match them peak and trough on some regular enough basis to mutually stabilize and even increase. (See fig. 5.1 on page 67.)

Incoherent waves, occurring when we are fearful or in pain, may be picked up by someone quite close to us physically but do not radiate out very far because they cannot synchronize or cohere with other waves. They remain localized because they cannot, by simple physical law, extend into that realm of coherent frequencies. In this way, incoherent frequencies block our interactions with the higher heart frequencies, resulting in isolation, emotional deprivation, or spiritual void.

Sexual reproduction, for instance, beyond the simplest life forms (and who is to say what goes on there?), produces powerful enough pleasure responses to assure the survival of every reproductive species. Apparently, the higher or more advanced the creature, the greater is the reward. Michel Odent's *The Scientification of Love* offers a brilliant analysis of the ramifications of this response. All forms of nurturing radiate coherent waves, and around the clock and all over the planet untold billions of creatures are engaged in that which is pleasurable and life-enhancing. Each such event kicks up a coherent positive energy, no matter how miniscule—and in the universal field of frequencies, size is of no consequence because it is simply not measurable by any of our physical yardsticks.

Our planet as a whole radiates (or should) an overarching positive energy. Thus when we laugh, the world really does laugh with us through perfectly sound principles of frequency resonance or meshing (entrainment, in Heartmath terms), and, conversely, when we weep we generally weep alone simply from lack of meshing of incoherent heart-mind frequencies. If, then, our general state is incoherent, no amount of sound and fury, effort, discipline, or storming the gates of heaven on our part can inform some higher power of our plight, for any such power would of necessity be coherent and no doubt nonlocalized. Our local incoherency can thus find no resonance—which leaves us stuck with our private negative affair. (And to him who has, more is given—whether what is had is positive or negative.)

Consider a copulating mouse, who sends its brief pulses of positive coherency out into the ether, where they cohere with and reinforce the positive peaks of you, me, and other creatures. Then consider something of the opposite: A mouse cousin being eaten by a bobcat sends a brief burst of negative incoherence (and who wouldn't?), but this incoherent wave cannot radiate far from the mouse (or perhaps his immediate kin in the vicinity) because incoherent waves can't mesh with coherent ones. The mouse's negative reaction may be very brief, however: research has shown that when a prey animal such as a rodent is seized by a devouring animal, nature releases opiates into the prey's brain, which, in effect, anaesthetize it and may in fact kick up the mouse's reward center in one last burst of mouselike joy.

In the great explorer David Livingston's African accounts we find his report of being seized by a huge lion, whose jaws actually met through Livingston's shoulder as the great beast began to shake him violently. Livingston knew full well his end had come. He reported, however, that at that instant he went into a sublime state of ecstatic bliss. His accompanying guides arrived just in the nick of time to save Livingston's life, and, like the discomfort in the shoulder that was with him the rest of his life, that ecstatic state never fully left him and confirmed his already firm spiritual anchor. Livingston's mind and heart were one, and heart takes care of its mind. Consider the possibility that had the lion actually finished off Livingston's earthly journey, his mind in bliss would likely have melded with a larger bliss. We aren't born to lose!

Meanwhile, to get back to our saga of bobcat and mouse, the bobcat, purring away in delight as he consumes his mousy lunch, sends positive signals, which, ironically, join on some level with those of cousin mouse in his happy copulations as well as those of you and me and whoever else might be experiencing an enticing event of whatever nature. Hunting animals show obvious delight and excitement in their occupation, just as grazing and browsing animals are delighted over a particularly tasty morsel of grass or bush and will search out that particular plant next time. This seems like simple survival, but it is more: The one consuming sends positive signals, the one being consumed sends a brief negative one. (Admittedly, we can't speak for grass or plants, though some have long claimed trees can communicate with each other, and as a child I surely talked to them and felt they communicated with me.)

Generally, then, the positive prevails, not just on an individual level, but also on the broad-spectrum level. Yet negatives can pile up as a field effect and, if picked up and reflected by enough negative minds, can grow not by coherence, but by creating more incoherent emotional reactions in others. A never-ending loop effect between negative field and receptive mind can then lock in. An individual is not aware of this field and couldn't seem to break out of it even if he or she were. The negative field or archetypal energy of a very dark sort that Carl Jung noted to be gathering in Europe in the nineteenth century broke out as the fascist phenomenon of the twentieth century, bringing the greatest pandemonium in history to that point. Again we see that a repeated action or behavior can become habitual and, eventually, instinctual, and this can all operate below, not just reason, but awareness itself.

Who, then, can break the negative loop once it has become entrenched? Wherever I go, I recommend Robert Sardello's books *Love and the World* and *Silence* and Childre and Martin's little book *The HeartMath Solution.* If we practice the solutions they offer for opening to the intelligence of heart until those practices become habit, eventually they will be nearly as instinctual as heart beat itself, and we will see the world anew.

10

IMPERATIVES IN CONFLICT

The previous chapter explored a conflict between biological and cultural imperatives, but an equally serious conflict arising between two biological imperatives, though indirectly a cultural effect, can also overshadow our actions throughout life.

Consider this scenario: A mother has been sent a priceless heirloom, a fragile Dresden figurine, and proudly places it on the coffee table. Her sixteen-month-old enters the room, spots that new, enticing object, and toddles over to investigate. One of the toddler's primary mammalian instincts literally drives him to interact with every unknown object in his environment—to look at it, taste it, touch it, smell it, talk to it, listen to it, and build what Jean Piaget calls a full sensory structure of knowledge or neural network for cognizing, recognizing, and categorizing the object or any like it.

The mother comes in, sees toddler making a beeline for this fragile piece, and calls out in alarm: "NO!—No, don't touch!" The toddler pauses at this intrusive command contradicting his internal command to investigate. He looks back to establish eye contact with the mother, for he has an equally powerful biological imperative to maintain contact with that caretaker, the one who gives life itself. Indeed, separation from this person is the greatest single fear of an infant-child's life, a built-in instinctive anxiety with its roots in real danger from saber-tooth and other such creatures.

As powerful as the word from this person is nature's other prime imperative: Investigate new objects. Further, any object or event in the nest (to the toddler, the nest, or home, is simply an extension of the caretaking mother) is fair game for investigation. The instinct for learning and establishing new relationships drives the ancient sensory-motor brain on in its investigation, while the child's emotional-relational intelligence drives him to maintain that critical relationship with the parent. With eyes locked on the mother's face and hands outstretched toward the figurine, he stumbles and gropes toward his goal on the table.

Later, when the mother (though it could as easily be the father) relates this story to her spouse, grandmother, neighbor, mother-in-law, she will say: "And that little devil looked me square in the eyes and did exactly what I told him not to do!" Repeats of this "deliberate disobedience" bring a variety of correctives, from mild to harsh. Above all we civilized parents are taught to believe that children must learn to behave. Modifying a child's behavior for its own sake, regardless of circumstance or logic, always takes precedence, and we enforce such learning "for the child's own good."

Natural inquisitiveness and investigative discovery learning is far more effective and permanent when a toddler is given a name for his exploration while the interaction takes place. Naming not only gives social sanction to the event, it also establishes such events as part of the parent world and involves the toddler's muscular language system involved in memory, all selectively building that child's own structure of world knowledge. A toddler expects a name for his explorations. "No, don't!" is not a name.

Allen Schore's book *Affect Regulation and the Origin of Self* (twelve years in the writing; 2,300 research citations) focuses on the toddler's emotional reactions to these negative correctives. The molecules of emotion involved play a critical role in all memory, learning, immunities, and health at this stage and are integral to all relationship throughout life. Schore found that the toddler's newly formed prefrontal cortex is the part of the brain most affected by such conflicting signals and the emotional turmoil they cause. According to neuroscientist Patricia Goldman Rakic, the prefrontal cortex is intended eventually to become the governor of our four-fold brain and its behaviors. Paul MacLean has called the prefrontals the "angel lobes" because they are the channels for compassion, love, care, and understanding, for nurturing and being nurtured. As we have learned, any early impairment of prefrontal development affects all aspects of development and relationships thereafter.

THE BRAIN'S DESIRE VERSUS ENCULTURATION

At each brain growth spurt—from development of the prefrontal cortex to the growth of the bridge, or orbito-frontal loop, between old mammalian emotional and ancient reptilian sensory-motor brains to the maturation of the prefrontals as the true governor's of the brain by age twenty-one or so—the fundamental question asked by nature is: can we go for the higher evolutionary intelligences now or must we defend ourselves or reinforce our defenses again?

We can examine once more the two imperatives that drive that young life toddling toward the Dresden figurine: The first is to explore the unknown world and build a structure of knowledge of it through both exploration and imitation of parent models and the cues they automatically send, and the second great imperative is to maintain contact with the caregiving, protective mother. Both of these biological imperatives will apply equal pressure in a more subtle yet far more expansive way in the budding adolescent.

In the home-nest all is fair game for exploration, but out in the wild beyond the nest the situation is different. Out there, such investigations of new phenomena are preceded by a check with the parent to determine the parent's general response. This parental check-in is as instinctual and critical to the infant-child as it is in all mammals—to determine the safety of investigation and also, in the human case, to get a name and identification of what we find. If the parent's response is positive, the toddler moves to interact with an unknown object on all sensory levels; if the parent's response is negative, our mammalian instinct rules for caution and even for ignoring the object or event. Instead the child may turn to other parentally sanctioned choices of the endless possibilities everywhere. Through this sanctioning response, the toddler builds a world shared with parents and others.

Whether indoors in the nest or out there, the child automatically picks up on subtle signals concerning the parent's emotional state through mirror neurons, the molecules of emotions, and other influences. If the parent registers alarm out there, the child will pick up on this alarm even if the parent doesn't express it consciously. If the child picks up an emotional "all's safe" sanction from the parent but the parent also issues an inhibiting "No, don't," blocking the child's action, the child is split between conflicting signals. As Bruno Bettelheim explained: You can't lie to a child because they are picking up more signals from you than you know you are sending. Or, as my meditation teacher Gurumayi once said: If, in your interactions with a child, that which you are thinking and feeling is different from what you are saying and doing, the child will be divided in itself accordingly.

Nature's model imperative is always at play. In every way, parent modeling enters into the picture and is a determinant in the nature of the structures of knowledge a child makes. Also significant is that if a parent's directives are ignored or continually defied by the child, punishment is used to emphasize the parent's determination that the toddler will learn to mind—for as we enculturated people learn early on, we must mind . . . or else!

THE DEVOLUTIONARY UNDOING OF THE EVOLUTIONARY LOOP

The shocking truth is what Allen Schore also reports: The average American toddler undergoes this confusing cross-signal blockage an average of every nine minutes—and the resulting emotional reaction in the child is far more intense and extensive than might be outwardly indicated. Full emotional-hormonal recovery from such an upheaval takes far longer than nine minutes, as the studies at the Institute of HeartMath have proved for years. (A minute or so of anger or fear can depress the immune system for hours before the parasympathetic nervous system can bring us back to balance.)

We are not aware of the ongoing chain reaction that such conflict sets up in the child (or ourselves), but it is surprisingly complex and critical. Schore devotes considerable research to this inner conflict that pulls the child in two diametrically opposite directions—both equally critical to survival, the classic double bind. Occurring on average every nine minutes, this conflict fills a large part of every day of a toddler's life and is always, of course, "for the child's own good." Our "No, don't!" reaction to a child's spontaneous exploration is as automatic as if he again and again reached out to touch a hot stove.

It is the critically important orbito-frontal loop, linking the prefrontal cortex with the other brain systems and with the cerebellum and reinforcing the links between brain and heart, that is the major victim of the barrage of behavioral modification every nine minutes. Between the twelfth and twentieth month of the toddler period, the orbito-frontal loop should be continually exercised as it organizes all systems into the single concentration of world-self exploration, and it should make its own imprint to the resulting patterns. From this exercise, it should also myelinate, making it a permanent part of brain structure. But the caregiver's incessant demand that the child modify his major exploratory actions according to parental notions of safety or social appropriateness begins literally to split that young mind to the point that a concentration of energies on the joyful task of world-making is fragmented and confused. Thus the tenuous, not-yet-myelinated neural connections of the orbito-frontal loop are not exercised and strengthened but instead begin to shift, disconnect, and reconnect in various detrimental ways. Those connecting loops linking prefrontals to the emotional-cognitive system are rerouted to form stronger and more powerful connections between emotional-cognitive and defense-survival brains. Once again, though nature has moved toward higher evolution, it has had to re-route and move instead for stronger defenses.

The result is that the child lives with one eye on the parent or authority figure (later teacher, neighbors, and so forth) lest they be displeased and scold, and the other on whatever object or event calls for attention. He can't concentrate his forces on a task at hand, whatever might need to be attended to and learned through forebrain response, because at least half his neural system is on the defensive. Unable to trust his world enough to allow full interaction with it, his attention divides, any focused learning becomes forced (the result of the command "Do this or else!") and stress increases, which throws even more emotional energy into the defensive hindbrain and leaves even less energy for real learning in the forebrain. A negative loop has formed, his focus is fragmented between two poles, he has no single vision that should fill his body with light.

In this way, the parental worldview that we unconsciously inflict on our child reflects the limitations and constraints to which we ourselves were subjected as toddlers. The enforced behavioral changes from our parents affected our own orbito-frontal loop and prefrontal development as we will, in turn, affect our children's. Our own automatic default drive demands conformity to social-cultural prerogatives, which are now focused on how we are molding our child. This assures a future citizen who conforms to those cultural directives—but if the child should not conform, we parents are ourselves held accountable: We are guilty in the eyes of our culture, censured subtly and indirectly or overtly in outright reprimands, even alienated and ostracized. None of us can quite stand such guilt because it is brought about by our own survival brain, and we feel the same ancient danger emanating from our old brain as we did in the implied or actual threat of abandonment that we suffered as children. Even though our parents have been replaced by culture, neighbor, in-laws, governments, ostracism—being cut off from the social herd—is the realization of the same abandonment threat. The amygdala doesn't register content or logic, only resonance.

We say and may really believe we act for a child's well-being, but our own parental well-being, social integrity, and acceptance are always there at a deeper and more pervasive level. The question "What will they think of me?" looms large beneath the surface. From whatever source, for whatever reason, parent-child bonding is blocked or compromised through these social correctives we feel impelled to induce. With pre-frontal employment impaired and the orbit-frontal loop redirected, the resulting deficits may show up in full-bloom years later, particularly in adolescence—and such are the ways in which culture replicates down through the ages.

THE PRICELESS HUG

Of course there are emergencies and occasions in which the child does need to obey us, and "No!" or "Don't!" must be employed. But Allen Schore gives us a remedial means of delivering the message in all cases.

At each occasion in which a negative-inhibitive corrective must be employed, follow it immediately, right on the spot, and preferably even as the correction is made, by reaffirming the bond. Simply pick the child up, hold him close in body-molding facial and heart contact, and hold that contact for a brief moment. You can afford that short time to save yourself and your child years of cumulative grief down the line.

This priceless hug is not to distract the child. An even greater advantage is gained by this simple gesture. Schore shows how negative corrections or constraints made in the bond bring positive learning: The bond makes the corrective acceptable and rewarding to the child. Jean Liedloff, author of the child development classic *The Continuum Concept,* claimed that the child very much wants to obey, comply, take part, and belong and he will modify his own behavior to do so, if the corrective is made within the bond. Obeying within the bond is a form of imitation that constantly reinforces the bond. Within the bond, learning and obeying are play and relationship—what life is all about. Through the bond, boundaries are set and happily accepted, for children seriously need boundaries by which they can define their world, selves, and relationships. As they grow, they move beyond previous boundaries and open to new ones fitting to each developmental stage.

If, however, we are more intent on punishment for misdoings or teaching the child to obey and mind—molding character, as the cultural myth calls it—those negatives will multiply and unfold for life, as we have seen. Beneath our behavior modifications there too often lies a rich vein of parental anger concerning life in general, and many of us seem all too willing to share this.

ONGOING BOND-BREAKING

Bonding with the earth unfolds early on but most prominently around age four, with the shift to the brain's right-hemisphere dominance. This bond to the earth is difficult to develop if the child is not first bonded to parents and free to explore that world. Making matters more difficult is the fact that such bonding is emphatically blocked or bypassed by virtual reality that takes the place of nature. This virtual reality begins with technological childbirth, continues through bottle-feeding, nurseries, cribs, playpen, television as babysitter in infancy and early childhood, Playstations, Game Boys, computers. Those not bonded with the earth will have bonded with an electronic world and later will support virtual reality at the expense of the living earth. The harm that results may not register in the child because ecological awareness has no receptors in a child of virtual reality. Further, any harm pointed out will be rationalized. Such a child can only seek ever more intense forms of artificial stimuli as he grows older, for he is dependent on these for maintaining full conscious awareness or feeling really alive.

Earlier, we discussed the extensive brain growth spurt around the sixth to seventh year, which results in a young brain gaining up to five or six times the neural-glial material for learning than he had at age three or four. What's more, this is five to seven times more neural capacity for learning than the child will have at age twelve, thirteen, or thereafter. The massive increase around age seven is lost by age twelve or so, literally pruned away by nature. Nature's dictum is "Use it or lose it." She offers an infinite expanse of possibilities through concrete operational thinking, but if the appropriate nurturing environment has not been provided and no appropriate models for such operations are given in those middle childhood years, such development cannot take place. Nature then responds as she must with a massive neural pruning somewhere around the twelfth year. She takes back, in effect, what was given and moves on in her developmental timetable. That window of opportunity that opens around age seven closes to varying extent around age twelve and the next window opens, ready or not. Just as twelve-year molars follow six-year molars, nature's plan is built around a general assumption of success, not failure; on her intent, not our negligence. Nature's plan is the only built-in framework, while ours must be arbitrarily superimposed.

Around age twelve, if the necessary steps of the ladder are there, higher realms of consciousness are ready to unfold, as development shifts from concrete operational thinking to formal operations, which are far more sophisticated procedures, allowing the mind to operate on its own brain functions and move into abstract thought, including creative ventures such as higher maths, philosophical systems—indeed, ranges of creativity far beyond the concretely embedded thought of the earlier period. Formal operational thinking is the first major stage, the launching pad for movement into higher worlds, which can lift us beyond all limitations and constraints of earlier life.

Formal operations can function, however, only by removing any unused portion of that remarkable allotment given around age seven. A mass of neural-glial aggregates going nowhere, using energy and producing nothing in turn, would crowd any new structures trying to form and drain the energy and attention needed for the far more advanced evolutionary systems opening at this time. Formal operations are so different from anything coming before that a fresh playing field is required at the same time that the brains utilizes all the previous gains and skills.

To accomplish this pruning, nature releases a hormone that dissolves all unmyelinated neurons and neural connections. Myelin, a fatty protein that speeds up signals from neuron to neuron and neural field to field, forms through repeated usage of a particular neural field—and myelin is impervious to the neural pruning chemicals released by the brain, so these fields remain intact. The unmyelinated, unused portion that is eliminated amounts to a staggering 80 percent of a child's neural system, just about the amount given in the growth spurt between ages six and seven. The remaining neural mass is the same contained in the cranium at age three or four and is the same the child will have as an adult.

This new creative possibility of formal operations seldom unfolds fully and sometimes not at all, for the concrete operational system that supports it largely goes undeveloped (hence the vast neural pruning). In failing to provide for concrete operations, we fail to provide the support for formal operations that follow. Even if a child is spared the virtual reality counterfeits that so thoroughly block development at the previous stages, he faces the massive hurdle of schooling, which subsumes the boundless, exciting voyage of discovery and learning that should take place between ages seven and fourteen or so, and instead reinforces enculturation as no other factor—not even religion—can. Culture and school, like culture and religion, go hand in hand. In the more than eight decades I have been around, school reform has been a constant chorus, about on a par with political reform. But reform of a profoundly wrong idea cannot bring its radical transformation. As a group of schoolteachers in New York State expressed years ago, we must simply burn the building down and let the ashes cool well before considering what kind of rebuilding we should undertake. Culture would of course prevent such a "burning," for schooling as it now operates supports our commercial culture too well.

So, in the period from age seven to eleven, nature has asked once more: Can we move on to the higher intelligences or must we defend ourselves again? And once more, she has had to move to defense, as she will almost inevitably have to at the next developmental stage.

ADOLESCENCE AND THE PREFRONTAL REVOLUTION

The neural pruning at age twelve or so clears the deck to make way for the third, most critical growth spurt a child has undergone since birth, a spurt that takes place around midadolescence. The prefrontal cortex and its silent companion, the ancient cerebellum, in the back of the hindbrain, are the arenas for this new and final growth, which lays down new neural tracks, as researchers put it, until around age twenty-one, in some cases even to twenty-five, if all goes well. If little or no preparatory development has taken place up to this time through appropriate nurturing, this growth will be so compromised as to be ineffectual and insignificant.

At this time, ideally the cerebellum undergoes a growth spurt not only as prefrontal support but to manage the astonishing body growth at puberty, with its overwhelming energy and the explosive emotional intensity of beginning sexuality. We speak of the clumsy adolescent who is clumsy for the same reason the toddler was in developing balance and walking: the teenager's growth is so rapid and extensive that the cerebellum can't keep up with it. It can't maintain coordination of those growing members and catch up on myelinating all that new growth and making movements smooth and effortless.

More important, as Rudolf Steiner observed long ago, the teenager is every bit as emotionally vulnerable as the toddler, for he is going through essentially the same emotional stage on a vastly broader level. He must now learn to relate not just to a new body and a new world and new forms of relationship itself, but to the universal consciousness itself that radiates out from the heart. At fifteen or so, he stands at that pivotal point on which our life centers and which aches in longing for its own evolutionary unfolding that is possible only through a unified brain-mind-body. And this unification takes place through the heart.

In his seventeen years with the Developmental Behavioral Biology Program of the National Institute of Child Health and Development (at the National Institutes of Health, or NIH), James Prescott conducted extensive research showing that cultural-social constraints and inhibitions of nature's biological plan are the breeding grounds for violence and failure to develop. Among the myriad issues involved are, as we've learned here, medical-technological childbirth and failure to breastfeed and provide sufficient body movement and contact as in carrying, rocking, holding, and general nurturing of the infant, toddler, and child. Just as important, he found that emotional support and allowance of sexual freedom for our young adolescent adults was equally vital for a peaceful society. (This explosive issue of sexual freedom was not kindly received—yet we willingly accept and react to violence with virtuous and righteous indignation . . . and more violence.)

Consider the requirements of nature for nurturing the toddler—protection, empathy, compassion, physical support, and a stimulating environment. As Prescott's studies show, these are as critical to the teenager, whose vulnerability and fragility is every bit as great. This required nurturing, care, and corresponding openness of opportunity is, however, diametrically opposed to the social-cultural situation the teenager faces today. He is, in fact, considered all but a public enemy and is treated like one, and he responds accordingly.

Remember that the young brain-mind is plastic, pliable, not fixed. This highly flexible youngster can be turned around at any time, if given support, models, and incentive. Most attempts to help are, however, ever more subtle attempts on our part to introduce behavior modifications for cultural conformity. The teenager picks up this underlying intent and feels all the more betrayed, for mind can't lie to heart. But he can be both turned around and lifted up if he's given the heartfelt support his own heart longs for. As nature's model imperative clearly shows, in a single encounter, one person can turn around any youngster—provided that person has him- or herself been turned around. Children become who we are, not who we tell them to be. Mirror neurons are never idle, and it is we adults who must change if we are to change the downward direction we have set for our young. Our species' fate is not written in the stars, but in our intent and resolve.

11

THE DEATH OF PLAY AND THE BIRTH OF RELIGION

In 1975, while working on my third book, *Magical Child,* a school teacher asked how he was supposed to teach children anything and prepare them to face the hard reality of life when all they wanted to do was play. I realized that all I had wanted to do as a child was play or be told stories (a form of play), and all my own children had wanted was to play or be told stories. I was working on the book's chapter on play and looked for some authority I could quote to defend or justify what so many saw as waste of time. I knew this teacher's "hard reality" was a fiction made real by eliminating play. (I remembered too well when I was six years old and being sent to what I thought was a nightmare world splitting me from the real world of play.)

I didn't find any satisfactory studies on the subject, and I pondered the issue for weeks, gathering more and more material, discussing the topic with workshop audiences and others. Eventually, someone sent me a paper by a Czech psychologist who seemed on the right track. Something clicked, and I felt on the brink of a genuine answer.

I gathered all my notes and tried to fit in this new piece, determined it held the key. I worked until past midnight to no avail. No matter how I maneuvered the material and pondered, all I ended up with was another rehash of some well-worn assumptions. Finally, exhausted, I leaned back, literally head-in-hands, and called out "Oh God, what is the importance of play in our life?" At which point a bolt of energy hit the soles of my feet and flashed through my body and I, with no customary transition of consciousness, went sailing out into space and felt myself flung from one end of the cosmos to the other in a sublime exhilaration, feeling rush after rush of ecstatic joy. Like a child being tossed through whole constellations again and again, I shouted over and over: "God is playing with me!" Finally the episode subsided, and I wept for some time, so emotional and sublime was the affair. (Much later I realized this episode followed Laski's Eureka! pattern.)

In regard to that chapter of my book, I still had no neat defense for play or even a good definition, but I never again doubted that play was the very reason for life, the reason why nature built such a lifelong compulsion into us to play every chance we get. Play is life living itself, nature celebrating herself, with no explanation or need for justification. We are born to play. The church comes along with some annual mea-culpa day of solemnity, sack-cloth and ashes, and we turn it into a carnival; the night before All Souls' Day becomes the celebration of Halloween, with its pranks and mischief; Christmas becomes Rudolph the Red-Nosed Reindeer and Jingle Bells; Easter becomes bunnies and new bonnets.

From the moment of conception, life expands through bonds of belonging, pleasure, and joy. Relating with each other and the earth is play. The more complex the organism, the higher its intelligence and the more complex its play of relationships, with the highest of all being the ever-unfolding expressions of love. Play expresses life's love of itself, the highest moral imperative.

Fred Donaldson and Stewart Brown show how animals play all the time—as long as no humans are around. Our presence seems to be a wet blanket on animal play; to catch them at it, we must either come upon them unawares or use long-distance telephoto cameras, as Stewart Brown did. (Donaldson takes a quite different tack, living with animals until he is accepted as one of them.) Brown has movies of interspecies play, with one of the most famous depicting a polar bear coming by regularly for a couple of weeks to play with an injured sled dog. Fred Donaldson teaches kindergarten teachers to play again so they can pass it along to children who too often don't know how to play, which is akin to a fish not knowing how to swim. Howard Gardner, Harvard psychologist and advocate of educational reform, once said the child never played with never learns to play. Donaldson has literally played with wild animals in Africa and with wild wolves and bears in different terrains. Michael Mendizza, whose Touch the Future Foundation is devoted to promoting play and nurturing of children, works with both Donaldson and Brown.

THE IMPORTANCE OF PLAY IN RELATIONSHIPS AND LEARNING

Early on, the infant plays with the nipple, plays peek-a-boo with that benign face up there—indeed, with great delight it plays at everything, knowing of nothing but play (until we set about to adjust it to reality and teach it to take life seriously). In play, every action is a learning, our brain free to imprint without censure or pressure. Under any such pressure, energy shifts back to the hindbrain, making difficult forebrain actions such as learning.

Working with mothers and newborns for decades at Case Western Reserve, Marshall Klaus said properly bonded and nurtured infants seldom if ever cry—crying is a primary mammalian separation call for the parent or for help, a signal that something is wrong. As the work of Marcel Geber, Mary Ainsworth, Jean Liedloff, Colin Turnbull, and others shows us, infants almost never cried in preliterate societies in which their mothers "wore" them in a sling and they and their primary caregiver were seldom separated. Mothers and infants bond in play—we need only listen to a mother's silly, high-pitched baby talk (produced by mothers worldwide, according to Alfred Tomatis's studies) and the infant's squeals of reflected pleasure. Primary mother-infant bonding assures affectionate-sexual love in the adolescent and adult. Loving, playful relationship that extends to society, the living earth, and creation itself is our natural state.

Playful relationship also establishes the context for all learning and development, and because learning and developing are the underlying survival instincts built into us, play is a survival instinct as well. This is what the infant, child, and adolescent wants to do and takes such joy in—if allowed. Through the ongoing mirroring of relationship within ever-widening matrices, children discover and define who they are and what their place in the world and, ultimately, the universe might be. Through the nature and quality of the mirrored relationship between individual and parent and then between individual and society is determined what is learned, which capacities are developed, and even what is remembered. As Michael Mendizza explains, what is actually learned at any time is the state in which the learning takes place—playful and joyful or grim and threatening.

Joy and pleasure are the bricks and mortar of physical, psychological, social, and spiritual development, and the developing brain must experience joy and pleasure if the complex integration of sensations is to take place. In those first three years or so (when the amygdala locks in its repertoire) an infant-child denied joy and pleasure, touching, caressing, and movement develops a brain that is "neurodissociative," as James Prescott explains, one that fragments rather than integrates experience. The same critically holds true at puberty and adolescence. Eliminate the safe space of pleasure and joy, acceptance and nurturing, and this expansive, integrated exploration of the world is curtailed and impeded. As a result, the adolescent regresses or dissociates into self-defensiveness, with its implicit violence that will finally surface as domination of others or be internalized as neurosis, illness, or suicide. The emotionally malnourished child may also experience an intensified sexuality at adolescence to compensate for an impoverished sensory-emotional system, but this sexuality is devoid of affection or love, is often violent and destructive, and is hardly conducive to family and nurturing of offspring, should any ever come to be.

THE ORIGINS OF RELIGION

A fundamentalist Christian, a seemingly bright woman with a Ph.D., explained to me at length that it was our duty as adults to break the will of infant-children, teaching them to obey in order that they, having then no will of their own, could be open to God's will and be obedient to him. If we fail to do this, she said seriously, and leave the child willful, both the child's soul and the adults' souls are imperiled and we face the risk of hell. So much for a loving God! Books urging corporal punishment from at least the fifteenth month on have been and still are bestsellers. "Spare the rod and spoil the child" has a long lineage, but we might change the phrase to "Spare the rod and spoil the corporate world that relies on such 'broken' children for laborers and our Pentagon that needs them as their fodder for wars." Without that rod, an actual individual might appear—a danger to culture itself.

A university professor I knew, a preacher in his early years before awakening to the travesty of his belief, wrote a brilliantly researched book on what he considered a major error of history: monotheism. (I remember our sixth grade teacher explaining to us that monotheism was the greatest realization our species ever had.) My friend's history showed that as long as any social group or individual could freely discover God within themselves and their world and work out whatever relationship seemed to flow, peace reigned. But when the notion of one God came along—a jealous, violent male God to boot—it wrecked everything, resulting in "Your God and mine can't both be the true one, so one of us has to go!" Thus the battle begins. On Ceylon (now Sri Lanka) Kataragama reigned within his twenty-eight-mile domain and he and his followers were happy and peaceful. Eventually, however, along comes the Muslim and his jealous God and all hell broke loose. How many centuries have Muslims and Jews been at each other's throats, playing eye-for-eye, with both sides blind as bats as a result? And they may well yet bring the world down, leaving us with the one true God: the God of absolute destruction.

A recent historian wrote that the language Jesus must have spoken was Aramaic, not Hebraic. The Greek translations of the Hebraic translations of the supposed Aramaic sayings of Jesus coming down to us in our gospels were made by the blatantly male-chauvinist followers of Paul the Apostle—and in every case of Jesus's reported use of the word for creator or creative process, these Greek translations use the word *father.* The actual Aramaic word used can in some cases be interpreted as gender-free but is generally feminine. Throughout history, that which brings about life, growth, harvest, that which gives birth of any sort is feminine, as is logical according to experience. Mother Nature is acknowledged worldwide, while Father Nature is apparently a Christian novelty.

A person's choice of metaphor will always, at some point, reflect his or her personal history, background, occupation, or chief interest. If we examine the rich metaphoric and analogic language of Jesus's teaching stories, parables, statements, and explanations of his cosmology or point of view, we find them all related to the earth and creation. The rhythms of nature and her seasons; plowing, planting, sowing, pruning, harvesting, rain falling on just and unjust alike—this earth and creation imagery goes on and on. If we read any of the stories or actions attributed to Jesus and substitute *she* or *creation* every time we run across that sanctified *Father,* an interesting light is thrown on the words.

The word *religion* comes from the Latin *religio,* meaning "to tie back, reconnect, relate again to something of the past." The origins of religion may lie in attempts to link again with the bond of life broken by culture. Yet linking to something in the past repeats the past in the present, as Robert Sardello makes clear. This blocks us from the present moment into which flows the future, when all newness takes place. It is here that we find the answer to the question of why we engage in violence and war.

THE DEMONIZING OF SEXUALITY

Religion as a foundation of culture plays its trump card by reversing our natural reward-threat systems with a demonic twist so that pain and suffering become virtues and physical-emotional pleasures become sins. Religion levies this brooding accusation and guilt against us all by, among many devious ruses, connecting this inversion to sexuality. Through this cultural ploy the human spirit diminishes while the religious pathology of guilt, sin, and hoped-for salvation (that is, being bailed out of an impasse from which we can't ourselves escape) becomes the foundation not just of culture but of our conscious mind itself.

For those who might question the logic in all of this, Paul the Apostle claimed that the more irrational and illogical his system, the greater the reward for those who could suspend their reasoning and nonetheless believe. Sam Harris's book *The End of Faith* explores this willingness to abandon the higher structures of evolutionary mind to revert to the lower, leading the author to brood over the "future of reason," which doesn't seem very bright.

In violation of evolution and countless millennia of social-biological heritage, the integrative nature of joy and pleasure (rewards) and the dissociative nature of pain (punishment) have been reversed to further control and modify behavior in ways opposite nature's imperative. Through fear and pain, outright physical punishment, or the even greater pain of abandonment and being cast out, religion wields its deadly sword. In infancy and childhood abandonment is the greatest fear and in the adolescent it indicates banishment from the gene pool and the pleasures and joys of sexuality and relationships in general, leaving the despair of isolation. As adults, alienation or ostracism from society and its joys of relationship is rationalized, compensated for, or even sublimated as we justify the "ways of nature" and our helplessness to change them.

Religious belief creates gender inequalities that strain the bonds of male-female dynamics on which family life and society depend. Whether Hebraic, Christian, Islamic, Buddhist, or Hindu, religion represents the creator-creative process in the paradoxical, irrational, and destructive image of a male God (the only male in history to give birth, however, metaphorically) who instills in us the intensely powerful energy of sexual drive only to then promise to punish us severely or even eternally if we give in to this drive in any way except through tightly controlled cultural channels of restraint. Checkmate!

Religion's deadly finger of guilt is levied on females as the originators of this sin of sexuality (turning on us males-as-victims with every movement). But it also touches male children as well, making sex or even bodily functions "dirty." Particularly in the Catholic tradition, any deviance from these cultural constraints concerning sex is considered especially sinful if we should enjoy the act. (If you must engage in such acts, however, better to marry than burn, as Paul the misogynist put it.)

The irony of these religious precepts is that the word *sin* originally meant "separation." "The wages of sin is death" reads the gospel injunction, and indeed separation can lead both to a barrage of afflictions that bring about disease and eventual death and to wars and other rage-expressive pastimes that can bring death on an ever greater scale.

How renunciation of sex became a religious issue isn't difficult to trace. In this we find the most efficient stranglehold on humanity ever achieved by culture: Religions teach that renunciation of body and its pleasures on behalf of spirit is the quickest route to salvation, which means deliverance from hell invented by a God of justice who was in turn invented by a rage-filled humankind. Even though we think we can reject religion itself, its dark precepts cast their shadow on all of us, bringing such guilt to sexual activity, for instance, that a host of troubles ensues: impotence or frigidity, resentment of the opposite sex, fear of parenthood, embarrassment over breastfeeding, alarm and embarrassment over masturbation in infants and children (in the nineteenth century, the sleeves of little children were tied to the bed lest they should touch their genitalia and find it pleasurable), and alarm over children's natural interest in the body of the other sex (or even their own bodies; Mormons used to bathe with their underwear on lest they cast lewd eyes on their own private parts).

This kind of lunacy haunts our lives. Such sexual blocks inflicted on us from infancy can become the very fabric of our self-sense, make breasts or penis signs of sin by default, and bring a general shame of our body's functions. At one period of the Inquisition and its fires, women wore thin lead plates to appear flat-chested lest some holy man's libido be stirred and she be made the next public barbecue. My Baptist grandmother (a young teenager when the War between the States ended) once stated at table that there was something vulgar about eating. Perhaps, I thought, we should eat in the dark as we copulate in the dark. The late George Jaidar, author of *The Soul: An Owner's Manual,* observed we should do better than that: copulate openly and eat in secret, thus doubly reversing and defeating our cultural imprints.

The so-called sexual freedom brought on in the revolt of the 1960s was short-lived before religion loosed its massive counterattack. And while the cork couldn't quite be put back in the bottle, this backlash made sure that any sexual freedom gained carried its price beneath the surface, bringing more guilt, negativity, and a riot of pornography or virtual sex.

In *The Biology of Transcendence,* I claimed that sexuality can be a key to transcendence, an opening to the higher worlds and a premonition of union with our Ground of Being, which can be a truly mystical journey between a spiritually bonded pair. (George Leonard, cofounder of Integral Transformative Practice [ITP] for realizing the potential of body, mind, soul, and heart, has written powerfully and beautifully on this spiritual aspect of sexuality.) Thus, of all our joyful expressions, sexuality is the one that must be curbed and crippled by culture, lest culture's bondage be cast aside. Culture brings endless ways of rejecting the body, leading to emotional and touch deprivation in infancy and childhood, particularly in the form of the inability or refusal to breastfeed infants. Breastfeeding is designed by nature to be a richly rewarding sensual experience for the infant-child and an equally rewarding sexual-sensual experience for the mother. As many mothers privately confess, orgasm is a frequent companion to nursing and a definite part of nature's plan.

Bondless and bottle-fed males (our overwhelming majority) are compulsively attracted to and fascinated by breasts throughout life, and while males are naturally attracted to breasts, the obsessive preoccupation that we see in our culture today was not and even today is not found in certain of the world's aboriginal societies, where the women go bare-breasted and use their breasts as nature intended—for nurturing children. In the corporate world, where pornography flourishes, breasts are used to sell goods, movies, television programs, and more, and women can be arrested for breastfeeding in public. One difficulty in reestablishing breastfeeding in America has been just this corporate ownership of breasts, which is a cultural ploy to use them to maintain culture coupled with women's discovery that through breasts they can wield at least some power in a male-dominated world. Thus the desire for large breasts, which has led to a new line of medical machinations that have made virtual breasts all the rage.

The joys and pleasures of relationship are obviously the glue that bonds and makes possible a true civilized society, while the virtual reality of electronics and medical prosthetics offer synthetic counterfeits that only further isolate and substitute for those needs they block and replace. Pleasure and joy are not only moral but are also morally necessary to develop a truly nonviolent, intelligent society. Pleasure and joy are literally the purpose of and prime impetus in the evolution of our cosmos, planet, species, and life itself.

Religion, however, in all its forms throughout all cultural history, has renounced the devil and all the desires or pleasures of flesh. Whatever the religion, body is bad and soul is good. Again we can turn to Blake for a response: It is "better to murder an infant in its cradle as nurse unfilled desire." Denying the body for the sake of the soul is the great trade-off with religion's God: Renounce the devil and all his works of the flesh—and above all, renounce the devil's consort, nature, the great (and of course feminine) enemy. We are urged to "struggle and wrest from nature her secrets and conquer her through them, bring her to her knees," as Bacon expressed it. Mathematician-philosopher Alfred North Whitehead claimed science could have arisen only in the Christian arena, but this renunciation of the flesh from which the nurturing of spirit all but disappeared is common to all religion.

The vilification of nature has taken many cultural forms beyond religion as well. Sigmund Freud, that enlightened guidepost of much twentieth century thought and neo-Darwinism, made such timeless declarations as: "[T]he principal task of civilization, its raison d'être, is to defend us against nature . . . [whose] elements . . . seem to mock at all human control; the earth which quakes and is torn apart and buries all human life and its works; water, which deluges and drowns everything in a turmoil; storms, which blow everything before it; diseases . . . only recently recognized as attacks by other organisms; and finally there is the painful riddle of death . . . nature rises up against us, majestic, cruel and inexorable. . . ." Interestingly, I have read that Freud reportedly had no sexual experience until age forty, when he married, and from this connubial couch he quickly retreated to then become the sexual authority for his time and indeed the last century. Yet more words from Freud: "Against the dreaded external world one can only defend oneself by some kind of turning away from it, if one is to solve the task by oneself. . . . [But] a better path: that of becoming a member of the human community and with the help of a technique guided by science, going over to the attack against nature and subjecting her to the human will. Then one is working with all for the good of all."

And we, caught in the guidance of this mighty enterprise of Freud's, spurred on by guilt, fear, and rage, destroy ourselves and watch as our living body of earth is destroyed, generation after generation. Now, at the apogee of this conquest of nature, we have the atomic bomb, the ultimate expression of hatred that is not only capable of subjecting nature to human will but of destroying her, once and for all. Soul, as a vehicle of religious hope, illusion breeding illusion, will then, I suppose, reign supreme in its rapture without the hindrance of body in any of its evil forms, even as the soul as our deepest core in the heart languishes.

"Goodbye, cruel world!" cries Freud and the fundamentalists waiting for the rapture—and "Goodbye, cruel man!" cries Sophia, that wounded soul of our dying world.

Part Three

The Rebirth of Spirit and the Resumption of Evolution

A holy man said to me, "Split the stick
And there is Jesus." When I split the stick
To the dark marrow and the splintery grain
I saw nothing that was not wood, nothing
That was not God, and I began to dream
How from the tree that stood between the rivers
Came Aaron's rod that crawled in front of Pharaoh,
And came the rod of Jesse flowering
In all the generations of the Kings,
And came the timbers of the second tree,
The sticks and yardarms of the holy three-
Masted vessel whereon the Son of Man
Hung between thieves, and came the crown of thorns,
The lance and ladder, when was shed that blood
Streamed in the grain of Adam's tainted seed.

HOWARD NEMEROV, *Runes* XI

INTRODUCTION TO PART THREE

The February 1967 edition of *Scientific American* had a pictorial essay by Damodar Kosambi concerning an episode he termed "Living Prehistory in India." An annual rite of sacrifice to the fertility gods to insure a fruitful, productive harvest was still being performed in a remote area in 1967 (though it may not be today) and was photographed in all its detail by Kosambi, who traced the origins of this practice to some two thousand years. Originally, a young man was selected a year ahead of the rite, declared to be the incarnation of the local god, inaugurated into this post through great ceremony, and then lavishly treated as the god for the entire year. He was given the best of everything, including young maidens eager to serve as his choice.

In another ceremony at the end of the year, anointed and bedecked, he was led to the fields made ready for planting. As part of the ceremony, a long pole or boom that could be maneuvered and swiveled and to which was attached a large iron hook and rope was mounted on an oxcart prepared with oxen in harness. At the appropriate moment during the chanting and ceremonial drumming, the great hook was plunged into the victim's back and the boom lifted up the bleeding body, systematically swinging it back and forth as the oxcart made the rounds of the fields to be planted. The flowing blood was thus distributed as an offering to the god until sunset, when the body was taken down, followed, before long, by the selection of the next year's resident god.

This had gone on for generations until one day a young man, exalted by his year-long position and the chanting, drums, and general excitement, went into an ecstatic state as he was led to the site, loudly proclaiming over and over that he was indeed the god! When the great hook was plunged in his flesh, no blood fell, nor did it all day as he was swung about the fields, arms outstretched, chanting that he was the god and blessed were his fields. When he was taken down and the giant hook was removed, it left no wound; no blood had been shed and the exalted victim was alive and well.

We can assume the crops did well, for from that day on no injury ever resulted from the ceremony. The position was hotly vied for by young men and the event still took place in 1967. The photographs accompanying the article graphically show the entire affair: the insertion of the great hook, taut flesh, body sailing about, the flesh whole and unwounded when the hook was removed at sundown. Even if the event was stamped out under the impact of technology, industrialization, and politics, Kosambi's pictures have memorialized it. Its meaning to us in this context: once the possibility of bloodlessness was planted in consciousness centuries ago, the possibility for variations on the effect was opened.

Jack Schwartz, a survivor of the concentration camps of World War II, displayed quite an array of paranormal capacities, including the ability to run knives through his arm, thrust large needles through his hands, and so on, to prove the concept of mind over matter, as it was called back then, and to point toward the depths of possibility within us. Schwartz withdrew from such public displays when he realized his motive was misinterpreted and he was treated as a sensation rather than as an example of what is possible. Although phenomena of this sort still occur right here in the United States today, with people discovering they can impale themselves in various ways, for instance, we hear little about such individuals, who walk softly and keep their heads down. Even more reason for this low profile: today, the big-stage New Age audience is currently caught up in health, healing, and healers as the box-office attraction, and any message is generally lost in the attractiveness of the messenger.

THE FAITH OF FIREWALKERS, OR GROWING A FIELD OF INFLUENCE

The following story, originally recounted in *The Atlantic* magazine in January 1957, is one I have retold many times, but I find that its message takes on new significance with each passing decade.

Leonard Feinberg, an exchange professor to Ceylon (now Sri Lanka) from the University of Chicago, observed and reported on a major religious ceremony held annually in the temple of the god Kataragama. Before the occasion, eighty candidates for the ceremony spent three weeks chanting Kataragama's name, meditating, fasting, praying, and observing celibacy, while the priests of the temple sprinkled them daily with holy water. As the preparation period neared its end, an enormous fire was lit. When it had burned down to white-hot coals, the participants were "seized" by the god Kataragama, who "breathed" them in long sighs, and they arose as a body and walked the pit of fire, with most of them remaining unharmed.

The fire bed was a recessed pit twenty feet long and six feet wide. Optical pyrometers registered the surface temperature there as 2,400 degrees Fahrenheit, which will melt aluminum on contact. In fact, onlookers could stand no closer to this pit than twenty feet because of the heat, and wads of paper thrown toward the pit burst into flame in mid-air. An average of 3 percent of the walkers died each time, though the rest, filled with Kataragama's spirit, were unharmed and exalted. The local explanation for the small margin of error was that the dead walkers' "faith in Kataragama did not hold." That average casualty rate could be considered a control group of sorts that proved, if we must have proof, that the fire was genuine.

At that time Kataragama was a viable, functional god whose field of power and intelligence was present and active throughout a twenty-eight mile radius of his temple. Feinberg reported various paranormal phenomena he witnessed within that area, most of which were rather like pranks played by a playful god, which was Kataragama's reputation. Kataragama's field of influence to bring about such phenomena was continually strengthened, built up over centuries by those who believed in him and by investments of energy, time, and risk of life. This phenomenon was a variation of Laski's field effect.

In our day, both some serious fire-walking and some superficial New Age gimmicks have occurred. Several years ago the truly spiritual Michael Sky led lengthy and intense fire-walking events that brought to a number of people true metanoia, or fundamental transformation of mind.

The accounts of Carlos Castaneda included local "god fields" throughout an area of Arizona and Mexico—all since dismissed through a thorough discounting of Castaneda, who had more to tell us than we were ready to hear. All gods are jealous, it is said, and a half-century later a fundamentalist scientism, technology, and religion have, simply through contact, destroyed most of these indigenous and independent fields and their respective gods, and the human powers and intelligences displayed by believers in these minor gods have mostly disappeared with them. Totally enraptured by and subject to our technology and its objects and things, we are not aware of what has been lost, so we instead doubt all of these paranormal accounts as nonsense lest our ideation be threatened.

Why is it first necessary to have belief in some abstract power out there in order to go beyond the ontological constructs of our world? Why is some capacity, such as a temporary immunity to fire or the ability to endure free of apparent wound a major puncture of flesh, attributed at first to some god-human interaction, though it then becomes available without the necessity of such otherworldly intervention? Perhaps this is because the god who appears to introduce us to some possibility within ourselves is within us all along, and the possibility offered within can change things without. Once knowing such change is possible, we don't need the inner prompt.

Neither god nor possibility is intended to be projected onto cloud nine and worshiped as the Golden Calf was worshiped but instead is intended to be incorporated into everyday life. If we project out the potential for any particular power, we may never find that such potential is within us. According to contemporary theology, Jesus tried to be transparent to his message, but theological overlay then made him the Christ and shuffled the projection a bit, which made Jesus a support of religion and culture and which ruled out the new mind, or Jesus's Eureka! revelation. As an analogy, suppose an event occurred on television that was plainly miraculous and could bail us out of our despair and we reacted by enshrining the television set on an altar and worshiping it. Mistaking medium and message can be a treacherous bog.

At stake here is that familiar model imperative of nature. Any potential within us that might be realized or manifested must be given a corresponding model in our outer world to initiate or jump-start the process. As we have seen in part 2, every aspect of the infant-child's development is dependent on this model imperative. But any new potential that has never before manifested, such as Gordon Gould's laser, must also receive a jump-start in some way—and, as in all such Eureka! experiences, this is accomplished through spirit. Gordon Gould's jump-start came from his own spirit, his own inner being (which is ours held in general), though this way of stating it would both embarrass Gould and offend the field in which he was involved. Can you imagine a scientific journal on optical physics running an editorial on the gift of Gould's spirit? Unfortunately, however, I have found no satisfactory substitute for the word *spirit* used this way.

Castaneda's don Juan finally tells Carlos that drugs and sorcery techniques were employed simply to get around Carlos's cultural imprint and open him to what was already within him. Contemporary science-technology has brought about serious doubt concerning any phenomena not quantifiable within the scientific frame of reference and therefore acceptable to scientific beliefs. Christian fundamentalists doubt for a similar set of reasons. All gods are jealous, whether scientific or ecclesiastical. The deadly effect of doubting such phenomena is their erasure culture by culture, across the globe—which also means such capacities have been erased from our own cultural mind-set. Doubt is the great enemy of spirit; it closes us in and shuts us down. Doubt of ourselves, humanity, and even life itself as fostered by science and capitalized on by religion, introduces to the planet a broad depression of spirit like that which has taken root all around us today.

We assume that technology and science are the highpoints of evolution. But evolution has been involved with the development of greater neural structures of the brain and the subsequent human capacity to move beyond the limitations and constraints of the lower animals. It has nothing to do with the objective, physical devices brains create for altering their environment. A rocket to the moon does not represent an evolutionary expansion. Reaching Steiner's higher worlds, Sardello's exploration-expansion of soul, a discovery of humans' immunity to fire and cold or our freedom from having to eat food—these are clear cases of evolutionary expansion. Our move beyond violence, war, and hatred would be a high-water mark of human evolution.

Meanwhile, every modern device we have invented is slowly (or with dizzying speed) changing the brain that is involved in the invention process itself. We are changing ourselves and destroying nature by the virtual reality we create. As both biologist Gregory Bateson and theologian James Carse as well as cellular biologists Williamson and Pearse have determined, mind and nature are a necessary unity; what we do is what nature is doing, and what we do to nature, we do to ourselves. It goes without saying, then, that just as we are an integral part of nature, nature is part of evolution, for evolution is a self-contained process accountable to nothing. Like space, it apparently goes on forever.

Evolution-creation is open-ended and subject to no laws or rules other than that which its own process brings about. Evolution can produce an infinite number of products or effects, possibilities explored automatically, which by default will include products that can loop back on and change the process itself in some particular instance of exploration. The evolutionary process can thereby produce a product that is destructive and that can self-destruct. This may happen within the cosmos all the time and may be happening here in our world.

Through a tenacious and open-minded interaction over a number of years, Dutch psychologist Robert Wolff found a wondrous world occupied by the Senoi people who lived nearly invisibly in the Malaysian rain forest. Not only did these people exist in a world closed to our five senses, but they also experienced an unbroken state of peace, love, and harmony with their world and each other. They had, as Wolff explained, no sign of the tension, anger, or depression that we take as a matter of course as part of our human nature and they possessed and needed virtually no objects of any kind. Each day unfolded a new adventure in discovering an ever-changing world of delight.

They had an edge on the Australian aborigines in that the Senoi could tune into and comprehend something of the ways and wiles of the white man who had invaded them. They certainly could tune into Wolff, who had to some extent tuned into them. Western thought was destroying the Senoi as a people by systematically destroying their rain forest to build rubber plantations. But the Senoi considered the trees sacred and communed with them. It was symbolic that they and their astonishing reality as reported by Wolff were dying off with their trees.

After several years of association with the Senoi, Wolff was casually and wordlessly initiated into their cosmology, their structures of knowledge, and he was thus able to discover at least something of the astonishing world the Senoi inhabited. This turned out to be the most profound experience of his life and an event that changed the country of his mind. He became aware of the profound setback in evolutionary development that Western humans and our Eastern counterparts have undergone. His conclusion and lament: we have no idea what we have lost.

We have no idea what states of being other civilizations—even those disappearing today—might have brought about or what they might have offered us. All we can see, grasp, or understand is the difference of physical products, the tangible objects and artifacts, the man-made stuff that different societies and cultures have left and how these compare to what our technology produces. We consider any society that has no technology as our evolutionary inferior.

Should we as a species become tone-deaf—that is, lose our capacity for tonal discrimination—we would be unable to perceive music as a sensory phenomenon or even comprehend the word *music.* It would be impossible for us to grasp that we had lost something if we had no neural system for experiencing that which was lost. We might at some point read of an ancient society that had once all but worshiped a phenomenon they called *music,* but we couldn't explain this phenomenon outside of its own parameters of sensation because it has no metaphoric equivalents. We can't say music is like anything. *Tone,* for instance, is what it is, not what it is like. And, for a tone-deaf species, *music* would be a useless, meaningless word without referent. If we follow this analogy, we might understand Robert Wolff's deep frustration at trying to get across to us what the Senoi had opened him to. Indeed we have no idea of what we have lost.

A society or race that has developed a brain system involved in states of consciousness might never be comprehended or even perceived by an object-oriented brain-mind capable only of re-creating objectified things and altering nature and, at the same time, knowing nothing of subjective internal states or experience. Such an object-oriented society might never know that some people might have nurtured and tended states of consciousness until they had evolved to astonishing heights and even become self-sustaining outside all physical aspects. Only a brain-mind that had likewise developed could comprehend and resonate with such beings.

Carlos Castaneda's don Juan referred to people who had built up such communally shared states and simply disappeared into them, so to speak, leaving no traces behind. Some people might scoff on hearing reports from Monroe Institute "graduates" discussing meeting in an "ethereal nonlocality." But those involved don't scoff; they experience. We have discounted experience on behalf of abstract concepts that create their own field and then determine our experience.

We can recall from part 2 that from age two or three to age six or so, the child develops internal imaging, which can then be projected onto external objects. At this stage, a matchbox might become a boat, a spool of thread, a car, and so on, with the child playing for hours in a world of his own making—divine play in its preliminary form. From around age six or seven to eleven or twelve, the child builds on this earlier play by developing concrete operational thinking, or the ability to take those internal images and actually change corresponding external objects in the world accordingly—that is, operate on an object with his abstract notions of what it might be and thus change that actual object. This is divine play on a higher level.

At around age twelve or so, the child enters the stage of formal operational thinking, in which he stands outside his brain itself and operates on the very possibilities of thinking and imagination, thereby moving into states of consciousness beyond concreteness. This higher level is itself only a preliminary exercise for the formal operation of creativity, which does not re-create in any way but instead originates states of consciousness outside the boundaries of matter entirely. We have no idea where this process of evolutionary development could lead because the entire evolutionary ladder by which we might grasp this potential has been truncated.

The Australian Aborigines developed these formal operations to an astonishing extent, encompassing capacities that we can in no way comprehend or duplicate (not even with drugs or with our virtual reality). They intuitively knew where all members of their tribe were at any time, though they might be separated by vast miles, and where in those desert wastes the underground water lay. They could detect a rainfall fifty miles away and move to intercept it, and they lived in perfect harmony and balance with that harsh land for near fifty thousand years. While the Aborigines were able to make some of the most sophisticated physical objects through concrete operations (such as the boomerang, which is a double hydrofoil wing assembly that can travel three hundred yards without varying up or down and, if it missed its target or prey, would return to the exact spot from which it was thrown), they kept their objects of possession to a bare minimum, emphasizing instead the state of mind that was their true treasure and mark of maturity. As a cultural practice, the Aborigines left no trace of themselves in their walkabouts, which, because they were hunter-gatherers, were constant and ongoing, believing the earth was sacred and should be disturbed as little as possible. Indeed, physical artifacts were rarely used because they were deemed a hindrance or encumbrance to their state of Dream Time in which they communed together and which they considered the real world.

Since we Westerners could not perceive the states of consciousness the Aborigines inhabited and saw these people instead as only naked savages, we considered them the lowest form of human life. In fact, we justified, they were too stupid even to build houses or make clothing. Dream Time, which apparently was far more rewarding than objects such as houses or clothing, was outside our perspective and world. There were five hundred thousand Aborigines on the continent of Australia when the white men first came with their guns and bibles, and by the late twentieth century, there were about five thousand left.

As developed, science is an expression of formal operational thinking limited to that which can be expressed physically, which means through concrete operational process. In scientific practice, then, the next higher rung of the evolutionary ladder, formal operations, finds expression through a form of concrete operations, a lower evolutionary ability. While this lifts the lower into the nature of the higher, as evolution is designed to do, the higher becomes entangled with the lower, as evidenced in our fascination with all manner of novelty products, which leads to the net devolutionary decline that is today destroying our earth. As a people stripped of all interior worlds, we have rapidly become dependent on the virtual reality we've created. Riveted on such a reality and its constant changes, we are changing with it. Science as an abstract application of concrete operations becomes a religion and a self-destructing process.

Cultural anthropologist Mircea Eliade spent some ten years writing his account of Tibetan monks (before they were functionally eliminated), studying and witnessing events that were incomprehensible to Western thought. A memorable item he related was the graduation exercises of an eighteen-year-old finishing his twelve-year training in yoga. He was required to sit jaybird naked on a frozen lake through the coldest night of winter and thaw by *tuomo,* or body heat, a stack of sheets that, though soaking wet at first, immediately froze solid. If he could not accomplish the task by morning, he did not graduate. As Eliade explained, where the faith is simple, the test is simple.

Before Eliade's time, the Englishwoman Madame David Neal, the first woman explorer to penetrate into Tibet, had written accounts of events she witnessed that were dismissed as fantasy. Particularly memorable was her account of creating a *tulpa,* an apparently living and breathing phantom person with a personality of its own. The ability to create this phantom had taken her six months of disciplined attention, but then, to the merriment of her sherpa bearers and helpers, she couldn't get rid of him and had to be bailed out by seasoned yogis who had warned her of the risk in creating these beings. She had concentrated on a wise and gentle sage, but the tulpa turned out to be a small, fat, treacherous little Chinaman who wouldn't be dismissed but claimed instead that he had every right to be there. (Tellingly, academics have long attributed such Himalayan accounts as Madame David Neal's to the high altitude that alters reason and makes individuals susceptible to illusion.)

Volumes have been written about capacities possessed by preliterate societies that are nonexistent today, even in the remnants of those people left (as with those sexually experimenting yet sterile adolescent Polynesians observed by Margaret Mead). To grasp what we have lost would require opening again to the potential of which those people were aware, which itself requires the mind-set we have lost through enculturation—a true double bind. But as we shall see, the mind-set of heart holds the way out of our cultural bounds.

12

LIFE'S STRANGE LOOPS OF MIND AND NATURE

We hear of the economy of nature, but nature operates equally by profusion, knowing no dearth of materials for making, for she owns the factory to throw out a million stars as easily as one. Why not, then, come up with that one star that sustains a planet that in turn gives rise to and sustains life? Why not a billion planets?

When I, a layman, consider DNA, I genuflect in awe. DNA and life seem to be a unity—a paired, reciprocal dynamic as in mind and nature, or creator and created. "Lift up the stone and I am there, break the stick and I am there"—wherever there is life there seems to be some form of DNA saying everywhere "I am here." Where does this double helix come from? It comes from life, just as life comes from DNA. Life is found deep in the ocean under incredible pressures, where an environment made by boiling thermal fissures provides for great slithery creatures, some as much as six feet long, that live and die like the rest of us. In gross polar extremity, life is also found in the form of algae locked in Antarctic ice.

A DIFFERENT KIND OF SIGHT

DNA produces photons, little bits of light, on a regular periodic basis, although that double helix is itself too narrow to be registered by a light wave (it is less than ten atoms across and is thus invisible to anything but an electron microscope, which, in turn, makes it visible to us). If each of the more than seventy-eight trillion cells in our body nurturing DNA produces a periodic photon, then a light body dwells within us. "If your eye be single, your whole body is filled with light," Jesus said. The DNA strand in each cell, though only ten atoms across, is about six feet in length. If you stretched out, end to end, all the DNA in one single body, that strand would stretch around the earth and to the moon and back several billion times. Looking within, we find the same eternity as that which stretches out.

As man is the measure of all things, we are, perhaps, the only limit with which we must contend. William Blake noted that the man who doesn't believe in miracles makes certain he will never experience one. Similarly, biologists Maturana and Varela wrote of the reciprocity of eye and brain: the eye sees what the brain is doing, and the brain does according to what the eye sees—a classic strange loop.

Our inner vision, which comes from the heart, can overlay our outer seeing of the world and give creative insight. To be chained to outer vision is to be sense-bound and constrained at every hand. It splits our vision from the unity of the heart, which prevents us from rising and going beyond limitation and constraint and closes us off from the possibilities of divine imagination.

Blake's challenge to us is to use our eyes in creative vision rather than using them merely as windows through which to see passively a fixed world devoid of spirit or life. Vision is a dynamic, not a one-way process. A failure of creative mind splits vision, resulting in our seeing one thing, feeling something else, and acting differently from either. Blake asks us to look evil squarely in the eye and see only the face of Jesus—for it is this act of creative vision or divine imagination that can give us a new situation.

THE IMPORTANCE OF POSITIVE COHERENCY

We can recall from part 2 that every positive, pleasurable experience we have sends forth a coherent wave of energy that can synchronize with other coherent waves and build a positive force field. The electromagnetic fields emanating from the heart reflect this coherence: positive signals from our emotional brain-heart interaction bring coherence between the first two wave fields of the heart's torus, which may open us to interaction with the third, universal realm, the Ground of Being, the realm of spirit and Vastness from which all newness arises.

Negative emotions, however, block interactions with the universal realm just as static can block a wavelength on the radio. Of course, the universal realm isn't effected in any way by this negativity, for it can't register negatives and so "never knows we are out of touch or in trouble," as Suzanne Segal said. Nor can that Vastness be informed of our trouble. While our body as a whole radiates varying forms of coherent positive energy, which is life-supporting, as we have learned, negative incoherent waves go no further than their immediate surrounding because they can't cohere with anything. The Neoplatonic philosopher Plotinus proposed that creation is the expression of love searching for its own reflection and fulfillment. If this is the case, to be sustained within our universe of coherent frequencies we call love, our world must radiate back out into that universe those very coherent frequencies. All is relationship; there are no one-way streets in nature. If we try to close ourselves off with a one-way street such as science or religion, then our resources dry up. Only in reciprocation and relationship do we find that our resources are renewable, whether individually or on the planet level.

Interestingly, a reciprocal energy exchange between earth and the cosmos may be undergoing alteration in our day because of the proliferation of electronics worldwide and the related flooding of the earth with the virtual reality of media. All forms of media, computers, televisions, and now cell phones (some equipped with televised visual stimuli) are everywhere in the world today. Media focuses on negative (at times violent) content, which, as it activates our startle-alert survival instincts below our awareness, grabs and holds our attention. Locked on to such potentially dangerous signals, our old sensory-motor brain is loathe to turn away from them, even though our mind, the last to be informed, might hate what we are watching. The overall effect of this worldwide negative energy nurtures culture even as we and our living world slowly succumb to depression. Meanwhile, the Vastness of the whole doesn't know that anything is wrong because it is a positive radiation outside all incoherent frequencies. Coherency and incoherency can't cohere.

COSMIC CHAOS, COSMIC ORDER

A recent satellite magnetic image of the earth's magnetic activity shows a remarkable mish-mash of chaotic pulses forming a deep web of confused circuitry on the surface of our earth (fig. 12.1). The image also shows that our earth still radiates those orderly torus fields of electromagnetic energy, as found around the sun in its vast multiplicity and in simpler form around our heart. If we look closely at this recent image, however, we can find no trace of any coherency in such a messy bag of electrical worms covering the surface of the planet.

Our evolutionary neural system was born from and geared for coherent electromagnetic fields in orderly harmony. Each of the myriad electromagnetic toruslike fields of the sun (see fig. 5.5) is, within itself, coherent and orderly, as are those large-scale fields of earth holding our planet's relation with the whole solar system. But the jumble of chaotic surface activity revealed by this satellite image is where we live and breathe, and our earth's protective belts can't protect us from our own man-made harmful radiations. Meanwhile, a strange planet-wide feeling of unease is slowly spreading, and the overall effect of this created chaos remains to be seen.

Our cosmos is geared for coherency within itself, and our life and brain-mind arose through harmonious relations within an orderly world. We are not built for chaos or disorder, but for lifting chaos into order. Coming into sync with our heart, which maintains within it the coherency of the cosmos, can lift us up and out of our predicament if we will take the time to tune in.

Figure 12.1. This satellite image shows the effects of all forms of microcircuitry on the planet, from the millions of cell phones, telephones, radios, and television sets broadcasting and receiving, relay stations, police calls, location finders, radar devices, microwaves, and so on. It has created a mass of static—not a coherent match in the bunch.

LAUGHING TOGETHER, WEEPING ALONE

So it used to be that when we laughed, our world laughed with us through perfectly sound principles of frequency resonance, and, conversely, when we wept, we wept alone simply from lack of coherent entrainment with a reciprocating energy. Suzanne Segal's claim that the Vastness doesn't know anything is wrong is a comment devastating to religions and their prayers. If our general state is incoherent, for whatever reason, no amount of sound and fury or effort on our part can inform some higher power of our plight. The cosmos, by its nature, is indifferent, and if we weep alone, we do so not because some stern judge in the clouds has not heard us or chooses to ignore us, but because the Vastness can't register our weepy frequencies.

As we "think" in our heart—coherently or not—so we are, and it may be getting more and more difficult to think positively in the heart. Despite our serious intent on a positive focus, energy tends to be drawn to defensive procedures below our awareness. If the way to get in touch with a higher power is through positive frequencies, which our depressed state lacks, we are, as we have seen, in an apparent double bind. While to him who has more and more is given, from him who hasn't more is taken away—even that little he has, our great model observed two millennia ago. This is true of individuals, nations, worlds. While our judgmental intellect cries "Unfair!" we note that we didn't create this world; we inherited it—though unknowingly we re-create it, moment by moment. Job's God would ask of us judgment-riddled intellects: "Where were you, when I laid out the corners of this Universe?" We didn't make the rules, but we are subject to them, and while with one hand we sow the seeds of havoc, with the other we plead piteously to be spared the reaping in store.

This brings us back to the issue raised in chapter 2 concerning the temporary state of worldwide shock following 9/11. As the world's most powerful nation and with the eyes of all trained on us more intently than ever, with the appropriate leadership we could have—for the first time in recorded history—changed the course of human events through forgiveness and reconciliation rather than retaliation and war. Our leaders, however, fed on that state of shock to loose a long-schemed web of insidious machinations for domination, while doing their dirty work by not only hiding behind the façade of waving flags, as is common for political tricksters, but also using the infinitely more noxious ploy of hiding behind the cross—the very cross on which was born into our world forgiveness as a way of deliverance from ages of strife and sorrow. The ongoing infamous evil in which our nation was then caught resurrected more powerfully than ever the dark demonic that has clouded culture's history during the past millennia. (In fact, when I saw our leader, in his bid for the vote, look straight into the camera and say "What you must realize is, I have been saved by Jesus Christ," I knew that the sad history of Christendom and its church that I had so loved had hit a low so ultimate and finally destructive that it had to go. It was at this point that I set about writing this book to do what I could toward that end. As Carl Jung once observed, an archetype can lose its usefulness and become destructive.)

The Roman Church symbolically left that body hanging on the cross, as is figuratively the case, for the evolutionary problems of an evolving brain-mind coming to a nexus there on Golgotha were left unresolved. These same problems plague us to this day and are again reaching toward a critical mass. Far greater trouble may be brewing there in the Middle East, ready to spew out on the world once more, as our next section will explore.

We join in that slain soldier's ghostly lament in A. E. Houseman's "A Shropshire Lad":

> *"God save the queen," we living sing . . .*
> *Oh God will save her, fear you not:*
> *Be you the men you've been,*
> *Get you the sons your fathers got,*
> *And God will save the queen.*

13

BRAIN CHANGE

We assume our current neuroscience offers a pattern by which we can interpret the brain-mind structures of ancient societies, but this may give only a very rough approximation at best. Brains change—which may help account for why culture has changed from a practical, commonsense practice of passing on knowledge gained by each generation to the instigator of the pervasive, deadly conflict we face today. But how do we escape the trap of culture's religions and rediscover the intelligence of heart and a life of spirit? To prescribe a remedy for our ailments, we must first have an accurate estimate of just what the ailment is.

THE SHUTTING DOWN OF THE BRAIN

New research on the brain shows its plasticity—its flexibility and ability to repair itself, even to create new neurons to replace lost ones—and also its fragility, it's subject to disruption of growth or function, particularly at certain critical stages. Neuroscientist Marian Diamond said our higher functions of brain will continue to shine brightly as long as they are challenged and used, but the higher newest brain in our system is the "laziest muscle in the body" and wants to go on the shelf early. Our ancient instinctive brains require far less energy and maintenance than the later evolutionary systems.

My physician friend Keith Buzzell pointed out that just as we build our brain from the earliest reptilian on up, adding the newer additions like layers on a cake, we sadly and too often die from the top down. Societies and cultures can undergo profound changes in overall brain development through changes in usage and application of those brain systems and can also die from the top down through disuse or misuse of them.

As we explored in part 2, the emotional state of a pregnant mother affects the nature, structure, size, and function of the infant brain forming in her womb. Consider the ramifications if her own awareness and emotional state are equally affected by her culture, which in turn is sustained and affected by the type of citizenry to which the culture gives rise. The strange loops turn again and again. Indeed, societies can go mad in spite of a strong core of sanity in them that tries to stem the tide, as exemplified by Nazi Germany in the twentieth century and the United States in the twenty-first.

In *The Biology of Transcendence* I gave a brief description of psychological surveys at Germany's Tübingen University covering a twenty-year period, from around 1975 to 1995, which revealed serious changes in mental states and functions in students. Notable shifts that were most prominent in those reports can be traced to changes in the reticular activating system, or RAS, of the sensory-motor reptilian brain. We can recall that the RAS is a gating mechanism that channels sensory information coming from the body into the higher brain structures (see chapter 9).

Our brain evolved in a low-intensity sensory environment, but today at birth and in the early years it is overwhelmed by our new electronic, high-density virtual reality overloads. The gating mechanism of our RAS selectively narrows in nature's attempt to screen out high-intensity stimuli that is inappropriate and can easily overload the young system. Milder forms of this closing up or shutting down take place in adults. As an example, workers in high-noise-density jobs tend to lose normal hearing. Different forms of desensitization take place in different technological situations.

Such effects in young people, as at Tübingen, have been written off as "positive evolutionary advances" that move us into brave new electronic worlds, but on examination, this rationale breaks down. Rather than some glorious evolving toward the stars or higher states, research shows specific devolutionary deterioration in brain growth and mental function from these changes. Those in charge of the study at Tübingen, for instance, observed that these young people tended to create an environment of high-density stimuli because, without it, they were subject to boredom bordering on anxiety. Their sensory gates screened so rigorously that ordinary or natural stimuli made little impression. The young people suffered a form of sensory isolation and anxiety when in quiet, nontechnological settings. Thus we hear boom-box thuds in passing automobiles and see ubiquitous carphones and headphones. Thus the nature of movies, computer games, popular music, and media in general grows in intensity to compensate for the growing insensitivity of the populace as a result of the higher gating of the RAS, which, in an ongoing cultural loop, is itself brought about by that high-intensive sensory level.

Recent studies in the United States have stated that the average six-month-old infant spends two hours a day in front of a television screen. TV is an excellent babysitter: a child won't move his eyes—or body—away from the screen. His visual system is literally held there by an amygdala-survival brain interlock, immobilized by stimuli from such constant change that there can form no stable imagery to which structures of knowledge can relate. This television exposure results in the primary sensory patterns of our reptilian brain locking into a danger-alert mode in reaction to the underlying flicker of the screen and the rapidly shifting imagery that create such sensory confusion that they indicate possible threat. This old-brain startle effect keeps the high brain alert for emergency and entrained with the lower brain as the child stares at the screen. Of course, the resulting immobility makes the infant-child "safe" so parents can neglect him.

Our primary visual brain evolved in an environment of stable imagery, and to develop this system a roughly similar environment must be provided. That mirror neuron-amygdala imprint functions from our beginning. Early developmental psychologists spoke of "face constancy" and the "stranger syndrome" occurring around the eighth month, provided the infant was in fairly constant contact with a stable caretaker. Around the ninth month after birth there formed "object constancy," a general visual stabilizing that indicated myelination of these primary visual components. All of these developmental guideposts previously observed have been challenged recently, however, which means nature's orderly plan for development itself may be breaking down or shifting and the old norms are becoming outmoded. We need only consider the remarkable change in the onset of puberty—down from an average age of fourteen or fifteen to eleven or twelve in less than a half century.

TV interactive games, computers, Playstations, Game Boys, and so on are bringing marked changes to the developing brain. Recent studies from Japanese scientists show strong neural connections forming between the hand and the visual cortex in children using these devices to any extent, along with a notable reduction in the development of the forebrain, the area connected to emotions and higher reasoning, language, and so on. These advanced neural areas are largely bypassed in the concentration on hand-eye involvement, and while these young people have lightning fast physical reflexes, they have little emotional control, easily become violent, and have serious learning difficulties. In his book *Hand and Brain,* neuroscientist Frank Wilson describes how these two evolved as an interactive unit. He emphasizes the importance of handling small objects to the early development of infant-children and praises Waldorf education for teaching kindergarteners and early-grade children intricate skills such as knitting, crochet, and other handwork.

Virtual reality gives counterfeit substitutes that work against development. It creates artificial environments much as the culturing of an organism in a test tube does. Entire societies or segments of them are undergoing serious change as a result, and the norms by which we can look objectively at our own situation may themselves be so altered that the old abnormal will become the new norm.

PERSPECTIVAL CHANGES IN HISTORY

Swiss philosopher Jean Gebser pointed out striking changes in visual perspective occurring in late-medieval Europe as reflected in pronounced changes in art, literature, and personal accounts of the time. Perspective of this nature involves not just visual spatial relations but also involves intricate patterns of neural organization of stimuli and shifts in the resulting conceptual-perceptual system. Gebser traces the impact this perspectival change had on both Europeans and those preliterate, non-perspectival societies encountered by Europeans at that time.

For example, he cites the Spanish conquest of the Central American Aztecs, whose available army of a hundred thousand blindly obedient warriors and blindly obeyed leaders all pretty much surrendered without question at the appearance of this handful of gold-lusting Conquistadors from Spain. In effect, they stood by and watched the Europeans rape and plunder their nation and send their gold back to Europe. Gebser attributes this collapse to the mind-set or perspectival view of those Spaniards with which the preperspectival mind-set of the Aztecs simply couldn't perceptually register or conceptually cope. The result for the Aztecs was massive confusion of mind and essentially mental-social collapse. (If we consider the wealth of claims of UFO abductees, we might, to be fair, apply this qualifying dynamic to their accounts: perhaps these people have experienced the extreme confusion of perspective an earth-reared, human brain-mind would inevitably suffer should it encounter a radically different frequency set found, say, in beings from another frequency or world.)

Julian Jaynes, a Princeton teaching professor, clinical psychologist, and author of the mid-1970s book *The Origins of Consciousness in the Breakdown of the Bicameral Mind,* uses the term *bicameral* or "two-chambered" in his title to refer to our right and left brain hemispheres. Apparently, these two neural structures functioned successfully in their division of labor until, perhaps millennia ago, they were derailed and began to break down to a varying extent over a wide scope of centuries. In the bicameral era Jaynes examines, ordinary exchange between the hemispheres had become dysfunctional. Before we dismiss this observation of neural breakdown, we might recall Allen Schore's description of neural disruptions in the link between the prefrontal cortex and the emotional brain in the early child and James Prescott's studies in developmental failures that bring about neurodissociative problems and other puzzling neural shifts indicated today. Jaynes's theory, derided by some academics, had both serious problems and valuable insights, one of those cases where we should have saved the baby rather than drowning it in its bathwater.

We can concentrate here on the positive aspects of Jaynes's argument rather than the troublesome word *consciousness* as he uses it in his title. In his limiting and constricting use of this word, Jaynes created serious semantic hurdles for himself at the very time the phenomenon of consciousness was beginning to be the focus of widespread and serious neuroscience investigations—and, of course, when the "new consciousness" movement was opening fully in New Age circles. Jaynes cut beyond all this fuss and offered clues to changes in thought and behavior occurring through alterations in brain structure itself. Such changes may have been brought about through alterations in cultural practices, as we find today, or may have been of an evolutionary nature (indicating that evolution is not through with us yet), or may have been the result of a geographical cataclysm of some kind.

Jaynes used the same historic episode of the Spanish Conquest discussed by Gebser as an example of a "post-bicameral" people—the Spanish, in whom an earlier bicameral breakdown had been repaired to some extent—unhinging the mental state of one of the lingering but decaying bicameral societies, the Aztecs. Jaynes argued that our present-day individual consciousness is emerging as a result of similar perspectival shifts in brain organization and function that started perhaps as early as three or four thousand years ago. According to Jaynes, the shift taking place was sporadic and haphazard, and it seems to me to be far from complete even today. Enmeshed in such a historical, possibly evolutionary change, we don't have enough perspective to know or recognize how incomplete our own perceptual-conceptual apparatus is, except to note our periodic warfare and largely illogical social movements of destruction.

THE SCIENCE BEHIND JAYNES'S THEORY

As a clinical psychologist, Jaynes worked with schizophrenic patients, and his bicameral theory of the brain stemmed from his study of schizophrenics suffering from mild to severe auditory hallucinations—uncontrolled voices from "out there" coming to them with intensity and virtually drowning out the world around them. On occasion, these voices drive their victims to terrible acts they would never have committed with a clear mind. Some 6 percent of the population suffers from this strange and terrible affliction.

Jaynes's work centered around the largely unanswered questions: Where do these voices come from, and why do they come? Jaynes's analysis revolved around the two hemispheres of our brain and the interaction between them. The early split-brain research of Roger Sperry, Michael Gazzaniga, and others (involving surgery to help epileptics, namely the severing of their corpus callosum) gave him valuable supportive data. The theory coming out of that split-brain surgery indicated that our right and left hemispheres each have a semi-independent personality or mind, which was an intriguing addition to previous proposals concerning the different functions the two hemispheres perform. As every school child knows, the left hemisphere—apparently connected with our aware social self, right-handed action, and superior ability to maneuver the body—is verbal, analytical, and intellectual. This left brain has fewer connections to the rest of the brain except for the late-forming corpus callosum that connects left and right hemispheres (and whose function is not wholly clear). This corpus callosum, which doesn't complete its own cellular growth until somewhere after age four, may have been a later evolutionary addition and is occasionally only rudimentary in some people. The autistic Kim Peak, for example, introduced in part 1, has no corpus callosum at all and a poorly developed left hemisphere. Peak had poor motor controls in certain areas, but tremendous pianistic abilities indicating sophisticated motor responses regarding music. (This anomaly calls to mind Julian Jaynes's observations about piano playing that opened chapter 4 and in itself offers clues to the riddle of mind and brain.)

The right hemisphere (with its clumsy, inept movements) is more holistic, with stronger links to the rest of the neural-heart system. While not dominant in personality and physical abilities and far less verbal, this right hemisphere is (as Elkhonon Goldberg points out) adept at learning brand-new material and quickly perceiving and interpreting what to do in novel situations, abilities that require fast associative thinking along with using the body's instincts and memory systems (giving us clues to the abilities of Kim Peak). In normal brains the right hemisphere automatically passes these critical learning-perception decisions to the physically dominant left hemisphere, where they are instantly acted upon by "right-hand" thinking and movement. This right brain gives the left brain new material to think about and, critically important to our survival, offers how to respond to novel or dangerous situations, moment by moment. Thus we, in our essentially left-brain awareness, require constant feedback from the older brains (reptilian and old mammalian), which we receive through the corpus callosum from the right brain (with its intact connections to those old brains). We are, of course, aware only of the final, integrated result, our commonsense response to daily life, but without this adaptive capacity given by the right brain, we would have only those survival instincts with which we were born and would never have survived as a species.

The left-hemisphere specialty of ours may be a later evolutionary development than the right hemisphere and its skills. The precursors for the left hemisphere, forming in utero as with all mammals, are slower in their formation than those of the right hemisphere, and while the prefrontal cortex completing them develops in both left and right from birth on, the left's overall development is considerably later than that of the right. (See figure 7.2, on page 105.)

Further, indications are that the left hemisphere, with its very sparse connections to the two animal brains, may have originally been connected to them as the right hemisphere still is. We might guess that nature eventually disconnected the left brain from those earlier systems in order to develop a new and higher intelligence, one not subject to the limitations and instinctual, reflexive reactions of the lower animal brains. This would open the way for language, abstract-creative thinking, and so on. Freed from our animal heritage, the human spirit and intellect could soar.

We could not survive with just this new left-brain ability, however, without the foundation of those earlier systems to draw on, a problem solved by the right hemisphere retaining its holistic stance and incorporating the lower animal brains into its service while keeping its ability to pass on information from the brain as a whole unit as needed by the new left-hemisphere operations. This involved increasing and developing that corpus callosum, which bridges the left and right hemispheres.

Bear in mind Karl Pribram's observations of the brain's overall unity and integrity: Any activity in one specialized area of brain peripherally involves the whole structure. All of this evolutionary biology may feed into Julian Jaynes's proposals and suggest that there was a long, interim period in our brain evolution when the left hemisphere was undergoing this transformation and was not very functional. During this time the right hemisphere had double-duty, which required arbitrarily signaling the left for actions needed. Skipping intricate steps in all the neural machinery involved, it's possible that the bicameral actions Jaynes suggested would have been the result: a left hemisphere that couldn't "think" for itself very well and was critically dependent on signals from the right brain on how to survive. In some regions of the world, the transition must have been early and smooth, and societies in these areas would move into a much higher state of being. The cooperative, benevolent, Edenic culture described by Riane Eisler as developing on Crete would require a balanced input from all neural-heart interactions. In other geographic areas where different environmental conditions and inherited practices reigned, the variations, partial adjustments, or outright dysfunctions might be extreme.

According to Jaynes, the automaton characteristic of our left-brain, right-hand capacity was the rule during this bicameral period. The left hemisphere, independent yet intellectually undeveloped, was rather helpless except for those signals from the right hemisphere, which the left instantly and reflexively carried out. Yet even these connecting links eventually broke down, according to Jaynes's theory, though perhaps that period of our history represents either the development of new capacities or the recouping or regaining of capacities lost during the transitional stage itself. Again, as Riane Eisler insists, at various times during history, humankind has functioned on a very high level of integration, wisdom, creative intellect, and harmony that have been periodically lost and then partially regained. In the period Jaynes analyzes, the corpus callosum required by these new capacities was perhaps as yet incomplete in its development, while the full development of left-hemisphere independent thought and a capacity for reasoning no longer subservient to the instinctive reaction of the lower animal brains had simply not yet developed.

In this scenario, an entire society might have functioned as a single unit of automaton thought, similar to that proposed for the Aztecs. Regardless, whatever the form of bicameral division of labor, it began to break down for any of a number of reasons. In his study, Jaynes examines those historic periods when this disruption happened at different times in different societies and over different geographic areas, creating in each case a social condition roughly similar to that which occurs when schizophrenia strikes a normal person today—something akin to a right-left hemisphere derailment.

The right hemisphere, with its holistic operation, eventually had to break into left hemisphere function as a strange voice commanding prescribed actions—much as Jaynes found in schizophrenics. To add more fuel to this notion, we can consider what happens in Laski's Eureka! experience such as occurred to Gordon Gould, William Hamilton, Kekule, and others. In these cases, mind (as an essentially left-brain action) is given a gift from the right hemisphere that the left couldn't achieve on its own. This gift requires the left brain to suspend its activity in order to allow the message to enter, so to speak. There may well have been many different phases of such a historic transition, episodes of clear left processing interrupted by the now older bicameral function breaking in, times of resurgence of bicameral thought and even conflicts between old and new, rather as we find in our adolescents today.

Again skipping technicalities here, Jaynes examines the early period in which this left-right exchange was taking place in the Mediterranean-Middle Eastern world, a time for which there is ample archaeological and historical evidence. The brain change seems to have spread slowly outward from here and no doubt occurred in other places around the world. During this shift, the right hemisphere spoke to the left hemisphere as auditory hallucination, according to Jaynes—not just as random thinking, as we might see today in roof-brain chatter, but in precise and commanding vocal form as the voice of an independent other . . . just as in schizophrenia.

At times of stress or unusual circumstances, schizophrenia seems to worsen in its victims. Jaynes notes how new and unknown conditions increased in those ancient bicameral breakdown periods, as when populations expanded from early extended families to ever-larger groups such as tribes, kingdoms, and so on. At each shift, general social situations changed or even deteriorated, calling for new responses and social controls. We can find similar situations today, with serious population overloads demanding serious shifts in behaviors, laws, governing systems, and so on. Through remarkably graphic archaeological remains, Jaynes traced this bicameral mind, with its right-to-left voice sounding as a command—generally threatening and demeaning—that brought about an automatic, unquestioned response from that physically dominant but mentally inept left hemisphere. Independence from the lower brains had not yet brought about independent thinking, leaving our dominant left hemisphere in a confused limbo.

This internal voice formed from birth and first came, quite possibly, from parental commands, as Jaynes found in the schizophrenics he studied. Jaynes showed how this voice of conscience grew from an individual, personal voice, to family, to large social groups and tribes with their taboos, to chieftains or shamans giving the commands, carried in a form of memory firing through the same amygdala-bound reflexes until the behavioral control of populations extended beyond all previous boundaries, leading to such situations as found in the Aztecs.

Other explanations are possible, of course, and may well work with Jaynes's model. In the German book *Vernetzte Intelligenz* (Networked Intelligence), von Grazyna Fosar and Franz Bludorf claim that "DNA gives a 'hyper-communication' outside ordinary time-space factors." They propose that communication between members of a species may take place in this way, and that possibly even prehumans knew such a form of group communication, lost in the evolutionary development of individuality, which certainly would fit in with Jaynes's general theory.

We have only to look at the examples of synchrony and moving as a unit found in herd, flock, or schools of fish. An even more astonishing aspect of hypercommunication can be found in insect hives. If we separate a queen ant from her hive, the workers continue working no matter how far removed the queen is. If we kill the queen at any point, however, then all worker activity stops. Such communication found in animals might have been lost as human individuality developed, but the DNA of a particular group functioning through some hypercommunication outside ordinary time-space may well have been at work in societies Jaynes has examined.

Fosar and Bludorf propose that "in earlier times, humanity, like animals, had strong connections with group consciousness and acted as a group . . . [whereas] to develop individuality, humans had to bypass hyper-communication. . . . [P]erhaps the capacity can be picked up again by DNA without loss of individuality, a way of accessing group consciousness that is not time-space bound." These research people suggest that a type of such group consciousness may be trying to form in what are called "new children" today, those youngsters who seem to have these kinds of connections with their group. Perhaps they represent a new possibility evolution is working out, and as such might furnish the needed model imperative to other developments. If this were the case, it might explain how, in an earlier period, capacities in the left hemisphere might have been first lost and then later picked up again once the brain's separation from the animal brains and new development was functional. Brains, after all, do change.

We return again to Laurens van der Post and the Kalahari !Kung and Robert Wolff's reports on the Malaysian Senoi and their intuitive capacities and gentle, benign ways. Such capacities involving top-down influences are aspects of that Darwin 2 effect discussed in part 1—an effect from which comes all our notions of God and personal demons.

CULTURE TAKES OVER

In Jaynes's chronology, the voices from the right brain to the left increased until finally, when population size became near unmanageable even to kings, the gods who stood behind and bolstered the actions of those kings were called upon. The history of the growth of such gods is a rich part of Jaynes's book. Fields form from ongoing, large-scale action by a society. Fields of action, force, different forms of intelligence and emotional-relations are always part of the scene and surely were in this earlier period.

Bas-reliefs of this preaxial period (somewhere before 500 B.C.E.) that have been found throughout the Middle East clearly show kings conferring with or being counseled by their ruling god. Jaynes notes that gradually a crisis arose, however, as the voice from the god slowly faded, indicating perhaps a further evolutionary shift of hemispheric interaction. The left hemisphere may have accelerated its move toward independent thought or perhaps it tried to become reestablished and was as yet not very adept at the job (much like the unnurtured infant-child's failure to establish the orbito-frontal loop sufficiently to do its job, thereby producing an evolutionary setback).

Jaynes's collection of bas-reliefs from this later period show the king addressing an empty throne of the god, who has fallen silent. (Oh God! Why have you forsaken us?) Evolution was slowly moving toward an independent intelligence that was no longer as dependent on the lower animal brains but instead was able to use them or call on them as needed.

During this bicameral period, nature compensated as best she could. Social groups were held together by a commonly accepted god-as-other voice perhaps in conjunction with a private god-voice within. Culture as a means of social control began to take over individual conscience, making the individual as subservient to the group as it was previously subservient to dictatorial voices.

Historians have noted that the civilizations of this period, particularly those of the Middle East and Mediterranean, were seriously affected by a series of natural disasters. A major eruption of a huge volcano on the island of Thera in the Mediterranean, subject of myth, may well have actually occurred. Some archaeologists theorize that a tsunami of gigantic proportions in that sea, created perhaps by such an eruption or explosion, sent huge waves crashing onto the coast, inundating the low-lying Middle Eastern areas and destroying life in an unimaginable proportion. This may have been the Great Flood of both biblical and Middle Eastern lore.

Appeals to those silent gods rose to serious clamor after these disasters, and there grew the idea of making sacrifices to appease them. Among surviving groups of those Middle Eastern areas hardest hit, trade-offs were negotiated with gods on a large scale. (Such bartering with fate, destiny, chance, or gods, an ancient human action, goes on in our own inner dialogue even now, as so many of us argue with our conscience.) Finally, this practice of sacrificing to angry gods became so firmly rooted in cultural imperatives that individuals offered up their own children to the altars of these gods, who were probably just as terrible and vengeful as the God of the Hebrew scriptures, eventually invented as sitting in judgment of both his own people and other gods, thus outdoing all previous gods in ferocity.

If all of this seems a bit far-fetched, we need only read the accounts of the structure of the society the Spanish found in America in more recent times. The Aztecs, capable of building enormous monuments, temples, and cities and of predicting the movements of the heavens, moon, and stars, seemed in the very midst of this feverish bicameral madness of coming apart and desperately trying to maintain their ideation. Estimates are that at one point, on the sacrificial altars at the top of their huge temples, as close to the heavens as they could get, hundreds of young men were offered up to the gods by Aztec priests. On those altars the priests daily cut out the hearts of dozens of young men, holding high those still-thumping organs for all to see. For a time, the victims were from neighboring groups and were captured by the army, but eventually, all sources depleted, the Aztecs turned to their own children. (Similarly, in the Middle East's bicameral madness ages ago, it is said that first-born sons were chosen for sacrifice.)

We might wonder how that vast Aztec society and those young men could have been so docilely led to their death at such a rate that some Aztec populations entirely disappeared. Lest we feel too smug, we should consider the ways in which we, as a nation and individually, perform in similar fashion our sacrifices to unseen forces or powers. Indeed, those Aztecs may have been just as capable of rationalization as we are today. The morphic fields of our modern times may drive us just as inexorably and blindly in novel forms of cultural imperatives cut from the same old cloth.

BLIND SACRIFICE, BLIND RESPONSE

Studies of the underlying motives and patterns of contemporary suicide bombers show to what extremities belief in rewards from the prevailing god can drive young people, particularly when coupled with an underlying social rage. That the families of suicide bombers often celebrate these deaths is not just propaganda, for ancient traditions claim such a death assures a place in Paradise, not only for the bomber, but also for his or her families. Also a factor is that male and female adolescent Muslims are rigorously separated, with the ever-present and real threat of humiliating, shameful death should anyone violate the established social taboos. Of course, those adolescents—particularly males—are driven by the same hormones that have driven young humans for eons. Muslim antagonism over the freedom exhibited by Western women thus in some way contributes to social rage.

In addition, most young Muslims have never heard anyone question their culture's sexual taboos, and in some Muslim cultures, many youths receive no education other than reading and memorizing the vast Quran, which does in fact promise Paradise to any who die in the name of Allah while putting infidels to the sword. (Of course, the Hebrew scriptures of Judeo-Christian heritage also tell of the rages of Jehovah; Deuteronomy, for instance, recounts a slaughter as grisly as any found in history—and all carried out on behalf of a jealous God.) Add this promise of Paradise to the oft-disseminated Muslim legend concerning the young vestal virgins that wait on the other side for those cultural-religious male martyrs who may earn their way to such a reward. (Christian martyrs, too, have died for heavenly rewards.) In this way, we can see that a complex web of cultural and religious issues is involved in the growing worldwide suicide bomber violence.

We in the West also respond blindly to violence, in knee-jerk reflex, in thrall of equally powerful cultural field effects that have also been built into us since conception. Buried within a vast labyrinth of cultural myth, legend, religion, and flag-waving beliefs, a subtly cloaked tradition underlies our upbringing just as it does that of Muslims. Through education, textbooks, career orientations, Sunday-school directives, Ph.D. aspirations, and so on, culturally dictated, corporate-driven, religiously revered, incestuously inbred mental fixations all function in us the way cultural imperatives always have—as well-rationalized and unconsciously accepted as ever.

OTHER BLIND CULTURAL ACTIONS

As an example, as we have seen, for the better part of a century our women and infants have been automatically subjected to technological hospital childbirths that do remarkable damage that has been both well-researched and published for more than fifty years. Why? Not only because hundreds of billions of dollars are at stake annually, but also because we, as cultural automatons to our own fields, are subject to any action contributing to culture's sustenance and power—such as technological birth and the cultural bonding surrounding it. In fact, a recent article called hospital birth the modern woman's cultural rite-of-passage, as if it were an almost inviolable sacred tradition. We carry out these actions just as blindly as the Aztecs did and with rationalizations that are just as elaborate.

Just as blindly, we send our children to schools that fail them despite the raft of studies and signs that tell us our schooling is so flawed: children's obvious signs of distress, poor health, violent tendencies, and suicides. And what of the painful, well-cloaked memories the majority of parents carry concerning their own childhood? We automatically repeat it in our children—"for their own good"—and the children, in turn, model their present and future behavior on us.

Our rulers send our young people out to kill each other by the millions—and we all but make saints of those rulers, build monuments to them and honor and obey them, not just because we know that if we don't "Or else" might fall on our heads, but because we have been so enculturated that most people can't think otherwise. Similarly, our children know that should they refuse to fight the wars set up for them, they would be shunned and shamed—fear of that is strangely stronger than fear of the possible slaughter awaiting them, as Susanne Langer pointed out years ago.

We return now to Julian Jaynes: he looked to the *Iliad,* with its strangely nonreasoning but passionate heroes, vanquishers and vanquished, moving through its pages like automatons, as clearly representative of the bicameral mind ruling societies and history.

On reading the *Iliad,* we are amazed that such fierce slaughter was accepted without question, with the perpetrators explaining that they were simply following the demands of their ruling god-voice. Such voices were apparently beyond censure and automatically accepted by all, in spite of the fact that they were wildly inconsistent, contradictory, and illogical. Such god-voices are as active today, only in new cloaks and guises. "What you must understand is I have been saved by Jesus Christ." "Support our troops!"

Jaynes noted that in the *Odyssey* (appearing perhaps a thousand years after the *Iliad,* though still attributed to Homer) a clearly individual but confused, ego-centered mind, so descriptive of modern man, wanders through random, meaningless tangles and endless traps and snares, perplexed by each new situation now that the sureness and certainty of bicameral thought, with its god-voices and definite action, is fading. Jaynes assumed that a truly individual mind was emerging, but in looking around us today, we can only conclude that it is surely not yet complete.

An individual mind has been evolution's intent all along, however, though its emergence has been caught up, time and again, in that strange tug of war between the prefrontal-heart movement and the amygdala-old brain survival reflexes, a true battle between the forces of light and dark, with poor humankind caught in the middle and losing out both ways. As always, however, our great nature has other ways available to it, other aces up her sleeve, and all is not lost, as we shall see.

14

VOICES IN THE WILDERNESS

The worship of God is: Honoring his gifts in other men, each according to his genius, and loving the greatest men best; those who envy or calumniate great men hate God, for there is no other God.

William Blake,
The Marriage of Heaven and Hell

The roots of a violent "dominator" culture are so obscure we can only infer their genesis. Later offshoots of these hypothetical roots are given by Riane Eisler, who compares a gentle, partnership model culture, as seems to have flourished in ancient Crete, with the invasion of the Kurgan in Eastern Europe and the Middle East. The Kurgan's roots are surely obscure, a wandering horde whose bloody invasions apparently produced shock waves in the cultural memory of Mediterranean and Middle Eastern survivors. Nonetheless, any survivors would have, from that point on, tended to breed ever-increasing generations of violence as the reptilian instincts, amygdala complex, mirror neurons, and model imperatives underlying human development took their toll.

If we tally up the studies of James Prescott, Alice Miller, Lloyd deMause, and Riane Eisler, to mention only a few, we might see how evolution's higher organizing forces bringing about humanity move to strengthen and renew the fragile hold love and altruism have in history. In this stochastic creation, the human has sunk to subhuman levels time and again, but those higher forces that formed us continually move for our well-being, in effect staging slow evolutionary comebacks at each of our backslides, though we pray for Cecil B. de Mille lightning-bolt deliverance from the heavens. (As William Blake says, "As the plow follows words, so God rewards prayers.") We can recall the anthropological-archaeological suggestions that the prefrontal cortex may have appeared and disappeared time and again before achieving some modicum of stability some forty or fifty thousand years ago.

Maria Montessori's sobering comment that a humankind abandoned in its earlier formative period becomes the greatest threat to its own survival was made years before Alice Miller's study showing how the Nazi Holocaust in Europe may well have arisen from just such lack of nurturing in childhood. Interestingly, the stereotypical Jewish mother may occasionally smother her young with her compulsive attentions, but in so doing, she also nurtures 30 percent of our Nobel laureates (though Jews comprise a mere 3 percent of our population) and a disproportionate percentage of our precocious artists, musicians, writers, and creative thinkers in general. Europe, it seems, might have used a dollop of such mothering during that late-nineteenth-century period in which anti-Semitism was infecting the soul of so large a segment of the sick-souled European populace.

Through Allan Schore's addition to Alice Miller's insights, we can examine the lifelong effects of child abuse recently coming to light in the parental history, pregnancy, birth, and early childhood of Saddam Hussein. As gruesome a story of child abuse as any on record, this started with an equal brutality levied toward Saddam's mother herself and her subsequent deranged state, all integral parts of the play. Add to this disaster the automatic mirror-neuron imprinting of young Saddam's ferocious uncles who undertook his tutelage and the replication of Saddam's attitudes and actions by his own sons, and we get a hint of how even today such a history and series of actions could multiply into the genesis of a Kurgan horde.

This book has had as its target what Rudolf Steiner referred to as the "mystery of Golgotha," for I have long sensed that in that event lay a key to the way out of our mirrored madness. So entangled are we in survival issues that we can't be fully aware of the part we play in them, though we might see the play of forces in projected form. Safely long-gone and mythologically overlaid, Golgotha, as symbolic of the struggle between culture's darkness and creation's light, might break through to us unawares, slip in under our protective radar, and show us how the major portion of the human condition and of our current dilemma is nothing new but instead has played out again and again.

THE HEBREWS AND JAYNES'S BICAMERAL MIND

Riane Eisler observed that remnants of that murderous Kurgan horde settled down in the Middle East many millennia ago, eventually giving rise to that group known as Semites, who, true to their heritage, were in constant territorial warfare with each other. The Semitic group that left the most intelligible records were the Hebrews, as detailed in the Judeo-Christian Hebrew scriptures addressed by Eisler in the second half of her study, *The Chalice and the Blade*. The history of this particular tribe, whose survival as a group amidst other Semitic tribes that rose and then disappeared in that semidesert land, centered on the common Semitic bonds of wrestling for territorial possessions on the one hand (and a bloody hand it was) and wrestling with the problem of God on the other—a paradox that is symbolic of our whole human experience.

Regardless of the problems presented in Julian Jaynes's bicameral breakdown theory, his broad sketch of the rise of gods in the stormy Middle East is both intriguing and enlightening. Putting together Jaynes's and Eisler's markedly different accounts, we can see two facets of the Kurgan-descended Hebrew people: a people who for centuries longed for a God of love and mercy and who wrote of this in as beautiful a literary form as any found in history, and a people who felt the chronic need to hold on to their lands. Having stolen land from other Semites through point of sword and slaughter, the Hebrews faced the ever-present pressure of stealing more and more or being stolen from, which required the populace's constant war footing. As paralleled for us in our modern, male-dominated capitalism, they unconsciously felt the ceaseless need for expansion by further conquests to maintain the cultural ideation established by the Kurgan long before.

As we find today, the common folk surely wearied from such constant bloodshed and the need to live "like armed crustaceans, eternally on guard against a world they could not trust," as Blake would describe such a mind-set. Yet Eisler's nurturing partnership culture was undoubtedly carried there beneath the surface, primarily by the feminine population that was long subservient to the masculine aberration of the dominator model.

The God of the Hebrews needed to be two-faced, Janus-like, one side cultural and the other spiritual, and indeed the Hebrews created just such a hybrid. In fact, the God they invented and experienced underwent as many transitions and changes as did the people longing for and experiencing that God. The result was a contradictory, ambiguous history of rancor that is representative of the whole human story. Regardless of the many faces of God emerging, however, that tribe's longing of heart was as effective a bond as were commandments and authoritative behavioral controls.

This Hebraic history stretched over centuries of a Middle Eastern world filled with sacrificial temples to unseen gods who changed continually as the brain-minds inventing them changed. Over the centuries, various Eureka! events from souls stirred by love and benevolence spurred the Hebraic quest in ever more passionate pursuit, building up a field of longing fed into by generations. "As a hart longs for water in the chase, my soul longs for Thee, O God," reads a psalm of those times, encapsulating and enshrining that longing of heart. This longing was capitalized on by political figures as devious and lying as any existing today, and they, too, betrayed the heart to the lusts of pocketbook and power while controlling the populace to keep the temple coffers and army ranks filled.

It was at this point that the abstract, mythical creation of Jehovah, who served the dominator rulers of the Hebrew tribe well enough, emerged as the focal point of that tribe and its history. Julian Jaynes found in this numerous examples of the breakdown of the bicameral mind and the slow struggle toward an individual mind. An analogy to this is the history of George Fox, founder of the Quakers, who broke from his culture's harsh directives and remained open to inner directives for which he would wait as long as necessary. When these openings presented themselves, he passed through them without hesitation—which all too often led him squarely into prison. As noted in chapter 13, Jaynes equated the situation of the Hebrews to that of the people of the *Iliad,* who blindly followed the voices of their inner gods that directed them in times of crisis or decision. Indeed, the heroes of that Greek world moved like automatons directed by those voices within. Jaynes's explanation for this was that the right brain was speaking in commanding voice to the left brain, which was unthinking but physically dominant and which automatically and unquestionably obeyed the intuitive right hemisphere. These may actually have been the mind's evolutionary forays as it tried to reestablish balance—such balance that may well have been displayed by George Fox.

This response may not have been as mechanical or unconscious in the ancient Hebrews as Jaynes assumed, however. Abraham, for instance, father of that Hebraic tribe, had quite a tussle with his conscience regarding the instructions of that inner voice that was the God of the whole tribe and to whom all were subservient in preservation of their cultural ideation against collapse and chaos. Abraham's example may have been typical of the God-human relationship in that bicameral time. As the Hebrew scriptures clearly show, the cohesion of the tribe depended on each citizen obeying that inner God, and breaking from this was no small matter. Where conscience ends and God begins may be a fine line—but only in retrospect. In Abraham's inner conflict the grounds for individuality and reason were being laid, and evolution was moving toward the goal of regaining balance in the system.

This right brain-mind/left brain-mind dialogue, with its right-hand and left-hand counterparts, involves a division of labor within us recognized throughout history and expressed in a variety of anatomical ways. Vedic tradition speaks of the "two hands of God" of which we are the expression. When the Norse god Odin, approaching the guardian of the spring of poetry and wisdom, asked for a drink from the spring, the guardian replied, "The price is your right eye." Jesus reportedly "sat at the right hand" of his God: He was the action hand that follows the dictates of the left. Should our right hand offend us, we should cut it off, and should our right eye offend us, we should pluck it out: These were Jesus's powerful figures of speech to stress the importance of that left-hand, right-brain balance without which we, the ego system within a left-hemisphere, right-hand logic, fall into chaos. Jesus also commented on our eye filling our body with light. Meister Eckhart echoed this, saying "without me, God is helpless" claiming not that he was God, but that he was God's manifestation, his mode of being without which God was not.

Julian Jaynes would say individuality was in the making during the axial age of Jesus, but also—and more to the point—is Jesus's admonition that this individualistic left-hemisphere, right-hand human ego of ours could easily get out of hand and throw us out of balance, thereby disrupting the appropriate left-right (God-human) dialogue and bringing, instead of wisdom and creativity, chaos and disaster.

Jerome Bruner's book *On Thinking: Essays for the Left Hand,* a scholarly study of these left-right dichotomies published in 1963, explores how even our language and its labels reflect ancient judgments inherent within us. Our left side, the *sinistra,* the dark interior of hidden and private parts of mind, is the intuitive, subtle, poetic wellspring of all creation, wisdom, and newness yet is left behind as the open, overt, and aggressive right thinking of a masculine mode dominates left-hand feminine expression.

Jesus and the Norse god of the spring of poetry and wisdom were not discussing *sacrifice* in its usual conventional sense, such as the bloody sacrifices the Aztecs or Middle Eastern cultures made to their gods, but as the mind's guarding against dominance of our system by that right hand. The word *sacrifice* means "making whole," and great models such as Jesus were pointing out the two modes of mind within us that must function as a unit. That left hemisphere may have collapsed into its dominator mode early on in our history, shutting out the right hemisphere's contributions and bringing ruin upon us—not quite what evolution had in mind.

At play here is a still more powerful issue: Neither left nor right brain was designed by evolution to be dominant, for dominance was designed to come from the heart. Yet the heart's directives and functions depend on a *balance* between left and right brains. We can return again to the Vedic conception cited above: God *is* the heart, his two hands our left and right hands and brains, his modes of action. This same idea can also be expressed as two generic trees in life's garden—one of life, the other of knowledge, with each tree giving rise to the other to serve the greater realm of heart, which gives rise to both.

As we discovered in part 2, present-day research shows the heart's neural links are primarily with the prefrontal cortex and the emotional brain—with which our right hemisphere is more tightly connected than our left. Our notion of ego in its pejorative sense describes the kind of individual mind resulting when our personal mind and its absorption in the outer world overtake us and dominate us, with the result being that the inner world of heart and spirit are compromised, ignored, or even lost—which again, isn't quite what evolution had or has in mind.

According to Jaynes, over time the god voices had fallen silent and decision began to fall to the left-hemisphered individual, but if the left hemisphere has not been well developed and is not in cooperative communication with the right, a dilemma develops. Evolution was nevertheless on the move in the Middle East and the silence of God spread throughout the region. This brought about or took place amidst great cultural-religious ferment, a general wail of anguished pleas and prayers that the silent God resume the conversation of old.

At some point in ancient history the notion had arisen of *sacrificing* to the gods, in the killing or offering sense, as an enticement or bribe to influence them, to soften their hearts that they might have mercy on the supplicants. Sacrifice in this sense amounts to bartering with God—"I'll give you this if you'll give me that" (a genuine left-brain logical equation). Gods demanding sacrifice soon began to dominate the Middle Eastern cultures and finally became a mainstay of cultural cohesion. Temples were erected for such sacrifice and occasionally became the homes of the very God being appeased, while the powerful priesthoods running the show and reaping the rewards were the central focus of a blindly obedient populace (again, we recall the Aztecs) or at least a populace cajoled and threatened into compliance, as is often the case today.

Over time, driven by the oracular voice of the priestly caste in control, people were compelled to offer up sacrifices to the prevailing God—"Do this or else!" In severe crises or cataclysms the sacrifices became more extreme until finally children, often first-born sons, were sacrificed (again, as occurred among the more recent Aztecs). In Israel, misfit outcasts cropped up to exhort the priests and people to change and won, at least in Israel, substitution of animals for children. Eventually, God spoke only to exceptional people (or those with enough chutzpah to claim such authority) or to the king or the priests running the temples and shrines set up to appease Gods and help the leaders' earthly finances.

In a strikingly beautiful and provocative book, *The Other Within: The Genius of Deformity in Myth, Culture, and Psyche,* Daniel Deardorff describes the liminal, marginalized cultural misfits, prophets, and ascetics, who risked their lives to make various decrees of their own, generally in direct contradiction to the dictates of the powers that be. In the axial period these decrees were passed on as from God, of course, as those of the priests and kings were—and addressed necessary behavioral changes the populace had to make and the punishments for failure to do so. (This is still with us. Today, God has more spokesmen than ever.) While these deviants threatened the stability of the priestly government, their activities also indicated moves toward an individual mind rather than the bicameral group-mind used by the priests to hold in thrall the religious cultures of the day. Throwbacks (such as Amos) threatened to break the fragile priestly-social balance of a semi-individual mind in the shaky process of stabilizing. It was not politically correct to seek, as Amos did, direct guidance from a God who had not only gone silent publicly, but whose name could not be spoken nor image graven because he was largely the private property of the state—that is, of priestly domain. On behalf of their vague abstractions and all-too-real need for cultural-social maintenance, the priesthoods in charge generally dispatched these prophetic deviants in due time.

All of this was in the very air of that tempest-filled period in the Middle East, a mental atmosphere breathed equally by those Hebrews and concentrated for them through that father of the tribe, Abraham. As conditions deteriorated, the nature of these sacrifices to God grew ever more extreme until the Hebrews imperiled their very children, leading to the idea of substituting animals (which may have proved more profitable for the priesthood—animals could be sold or eaten and infants could not). In Abraham we find a powerful myth of this central archetypal action: The willingness of a father to sacrifice his own son in blind obedience to that inner voice of the tribe's God, an obedience critical to the sustenance of that tribal culture. (Later this was echoed in the sentiment of Caiphas, high priest at the time of Jesus's arrest: it is expedient that one man should die, lest the whole nation perish.)

Following great anguish of mind, Abraham, preparing nevertheless to carry out the terrible deed, neatly bypassed the whole issue at the last moment when a ram that just happened to be passing was immobilized in a thicket by its horns, thereby offering itself as substitute obviously sent by the very God that waited to be appeased. Along with becoming one of the great classical examples of faith conquering all, even as it clearly illustrates Julian Jaynes's bicameral mind obeying the dictates of that inner voice of the tribal God, this episode points to the rationalizing cleverness of Abraham as an early sign of the reassertion of a truly rational mind in the midst of that chaotic and decaying bicameral scene.

THE FIELD EFFECT THAT PRODUCED JESUS

Meanwhile, like a thundercloud attracting the energy of many disparate cloud groups, humankind's longing for a God of love grew in ancient Israel. This longing of heart built up into a field effect of serious proportions, and, as with all natural phenomena, reached a point of saturation so that it sought expression. In typical Laski fashion, a corresponding resonant target for that longing's expression was selected from among the people of that field's origins. Spurred on by that field effect itself, in the typical creator-created strange-loop dynamic, the field of longing and target attracted each other.

The religious establishment might be offended by a claim that the process of field effect giving rise to Gordon Gould's Eureka! is the same one that gave rise to Jesus. Yet while the fields involved were different, the same creative process generated the same play of lightning. In the case of Jesus, the answer given was cosmological or ontological, having to do with culture and our primary dysfunction, focusing on the very process behind this Eureka! phenomenon itself. Jesus described the strange-loop process giving rise to us and to which we in turn give rise, however unconsciously.

Theologians refer to Jesus as being transparent to his message, which is accurate, though they fail to see what that message was. It wasn't one of cloud nine, angels, and harps or those adulterating or distorting Pauline additions, but one of the creative process of which Jesus was and we are an integral part. His message was metanoia and the process behind his own Eureka!, and he and his translation were products of that process. If we understand the process of which we are part, the possession of the product is no longer such a passion.

Cloud nine and angels were a much later overlay on Jesus's cosmological-ontological insights. A variety of political-religious notions were added from other religious and apocalyptic groups, and ongoing textual alterations to various sayings and actions of Jesus eventually made up an accepted gospel long after his death, as Princeton's theological and historical scholar Elaine Pagels describes.

In Jesus, the target of this long-building field of longing was a liminal, semioutcast Israelite youth wandering in the desert of his mind. No well-oiled, well-heeled, learned priests in the temple, overflowing with their own knowledge, importance, and convictions, would fill the bill as target for the event moving here. What was needed was a hunger of soul verging on the desperate or reckless, someone willing to throw him- or herself away, in effect, to plumb to the depths of self and being without concern for self.

Jesus was a product of his culture yet did not fit its pattern. Otherwise, he could not have been open to that which was radically other than that culture itself. As a product of his culture, he brought to completion a major aspect of that culture's field of longing, which was centuries in the making, just as Mozart or Bach were products of the rich field of music inherent in their day, or Gordon Gould was a product of the field of physics in his. The lightning bolt that Jesus experienced generated out of not only the long quest of those Semites but also the general quest of that historic period as a whole. The great beings who have cropped up in various areas and times throughout history have focused on serious ontological issues, not on local politics.

Over time, then, these Hebrews, perhaps inadvertently, had produced the right circumstances that led the way to the evolutionary breakthrough that tried to take place in the events culminating on that lonely "hill of the skull." The word *tried* is key, for we must remember this is a stochastic creator-created world, which is to say Golgotha failed in any direct effect, though it has not yet failed completely in a long-range one. The wheels of evolution can grind slowly.

THE AXIAL PERIOD

Theologians refer to the period of history running from about 500 B.C.E. (B.C.) to 600 or so C.E. (A.D.) as the axial period. The word *axial* comes from *axis* and refers to the proposed point around which our evolutionary history revolved, namely B.C. (Before Christ) and A.D. (Anno Domini, or year of our Lord or After Divinity)—Christian conceits to say the least but true enough symbolically. This period of time is equivalent to that of about 900 C.E. to the twenty-first century, quite a stretch of time by any reckoning. During this long millennia, a series of great beings appeared, chief of which, in order of their appearance in our own play of history, would probably include Lao Tzu, Buddha, Krishna, Jesus, and Muhammad. Through copious archaeological and anthropological research into these figures, scholars can tell us quite a bit about this historical period, although insight into these actual people themselves remains fairly obscure.

These figures played a key role in and are themselves exemplars of the changes of cultural mind-set that took place in their time and with which we are still very much entangled. Underlying this axial millennia and its legendary figures is a central theme that is theological, cosmological, psychological, or historical—however you choose to describe it—but which in Vedic tradition was termed the struggle of God to rescue his bride, the soul of the earth and heart of humankind, from the forces of darkness. This struggle is hardly over, obviously, in spite of the efforts of the various champions serving God in that struggle. A serious evolutionary shift of brain structure is involved in such a "rescue," and time, as we've seen, is inconsequential to evolutionary cycles.

In looking closely at the various influences involved in the Golgotha event, we find that Darwin 2 evolutionary movement of higher forces—the movement that had brought humanity into being—occurring this time to *restore* humanity to balance. At play were the prefrontal-emotional heart unit on the one hand, and the amygdala complex and ancient survival instincts of the reptilian brain on the other, which is to say the evolutionary tussle that ensued was biological, with *biology* here again meaning the logic, order, coherence, or structure of this life system.

MYTHOLOGIZING GREAT BEINGS

We know that each of these axial figures had the historical basis of an actual person who had amazed his friends and confounded his enemies sufficiently to furnish the source of what the cultural anthropologist Mircea Eliade defines as the nucleus for mythological overlay. Eliade points out that mythical overlays of any power and dimension aren't built on lesser figures or idle dreamers. Only genuine movers and shakers, whether or not we care for the nature of their moves or shakes, capture the imagination and qualify as targets for such historic movements of mind.

Once the process of mythological overlay begins, that original figure is fair game for the plays of mind made by anyone attracted to or resonating with that figure, even if the attraction is negative. Through this attraction, whether positive or negative, an individual adds to this mythologized figure his or her own bit of interpretive or imaginative insight. It is this very interpretive overlay and the compulsion that engendered it that can become the real value of the original figure.

Through mythical overlay, the kernel or seed of origin of such figures grows like the layers an oyster grows over an irritating grain of sand. And these mythical figures, however varied their linguistic cloaks, were probably real cultural irritants in their day. Nevertheless, most of them finally emerged as genuine pearls of great price—a price that all too often each personally paid. Over time, these legendary figures with their accumulated layers of belief, theories, and passions are bartered, bought, sold, lied about, and warred over until the original person underlying them may be long forgotten.

A recent article proposed that the figure of Buddha, for instance, was almost surely a composite of any number of exceptional Hindu saints who percolated out of a different pot than the conventional Vedic-Upanishanic brew into which they were born. They may have been a motley crew of social misfits scattered over generations, perhaps, but gradually attracted and were gathered together through the field effect of generations of seekers and followers. The scattered reports and tales these followers left behind were assembled and reassembled again and again into a collection by their interpreters and promoters, and what results is the rubric of a final symbolic figure so large and impressive that it gives a stable and attractive image we can hold in our mind's eye.

If we delve into such mythical beginnings, we generally find a fictionally reconstructed childhood and romantic history by which we can anchor that figure in our own emotional memory and time-space. In Buddha's case, we have a youth who renounced the wealth of the world and subjected himself to the desert of a twenty-year sentence of sitting beneath a banyan tree, with its great, hovering serpent, waiting patiently for a well-seasoned and tempered enlightenment. (Marghanita Laski might ask: after such enormous effort, what else would you expect?)

Generation after generation, subsequent disciples pick up and focus on the wisdom words and phrases that have stuck in cultural memory and are relevant to our ever-changing times. From these grow ever more layers on the pearl. The question thus arises: Will the real Buddha please stand up? This can be a futile question, however, even if the answer could conceivably be unearthed, for the original figure might be quite insignificant—even of little worth—once stripped of the finery added over the centuries. The worth of a mythical figure may well lie in the resonance that attracts, fosters, and nurtures our febrile imaginations, allowing us to add to or change the figure's outer garments generation after generation as needed to meet the spiritual hungers of our day. Like any field effect, myth is sustained by reciprocal loops between field and mind, and our investment in such a field enters as an interdependent, reciprocal, and necessary part of what we receive from it. We sow and reap our gods accordingly.

The idea of stripping away overlays of this sort to get to the real McCoy can therefore kill the effectiveness of any such figure. Ancient Sanskrit scholars tried to preserve the purity of the Sanskrit language by freezing it into complex syntactical laws that had to be learned if anyone was to read Sanskrit or become a Sanskrit scholar—no easy task. Bound by such restrictions, it wasn't allowed to change according to cultural shifts and needs, and Sanskrit eventually became a dead language. Essentially, the same happened to the less rigid classical Latin, preserved in any useful but very faint fashion only in the flexible Romance (once Roman) tongues to which it gave birth. We can also try reading the old English that existed before Chaucer's time, or even the original middle

English of Chaucer to find yet another example. Chaucer, however, was instrumental in shaping the English we know today, for his language was a flexible enough invention to survive the centuries and it is still dynamic, changing even now around us. Unless a language can change it dies, no matter how the new lingo generally irritates us old-timers (who may have frozen into a fixed-brain position).

Unless our great beings of history can somehow remain fairly unbound by tradition, they too lose their effectiveness—and we serve a dead tradition. Each disciple or scholar investing in his or her interpretations and translations of such historical figures tends, however, to establish boundaries around his or her own interpretations and then defend those boundaries fiercely. These individuals might then look for their own disciples and establish their own tradition. Lutheranism, Calvinism, Catholicism, Methodism, Muhammadism, Buddhism, Hinduism—all are variations.

The word *tradition* comes from the Latin *traducare*, "to traduce," "to betray," to trade on or barter away. Tradition is a snare, locking the present into a replication of the past, a dynamic similar to the one that led to the fate of Sanskrit, with its vitality diminished and its fresh content immobilized in formal overlay until it had gone stale and was spoiled and of no use—except to those diehards upholding the tradition, those whose identities are locked into it, whose egos are invested in it, or whose fortunes depend on it. Tradition and religion go hand in hand—two cultural supports tying us to replications of the past and blocking the unfolding of our future in the present.

The translators of such many-layered original figures dream up new overlays that determine the ongoing worth of that growing pearl, even as they may further obscure the genesis of the whole operation. In this way we arrive at Chuang Tzu, who translated and thus contributed to the mythology of the mystical Lao Tzu, giving an effective and brilliant interpretation as relevant to our day as back then. Next, however, came Confucius, who acted to the Taoist tradition as Paul did to the following of Jesus, building a counterfeit offshoot different from the original and deflecting people away from the genesis while selling well on the political-religious market.

Buddha had similar followers and translators, an ongoing host of such figures, still active today, vying with each other for attention, each claiming to have the Truth. My friend the late George Jaidar once proposed that our inherited accounts of Jesus, which was a popular male name of the time and shared by many, may have been composites, several characters assembled into a single synthesis in the short two or three decades after Jesus's death. Israel had many revolutionary firebrands at the time, running about stirring up trouble and getting into trouble themselves.

Although I couldn't buy into his proposal when it came to Jesus, according to Jaidar, such composite myth-making might account for the serious contradictions in the actions and sayings of Jesus handed down to us and the far more serious contradictions by those acting and writing in his name. Compiled over many generations, each editor deleted, added according to taste, or simply rewrote the whole Jesus affair in new settings, which is the way mythologizing great beings goes, an invaluable and continual process. Today, among the bestselling items in the book world are the varied accounts of Jesus and Mary Magdalene. (Years ago James Carse, theologian at New York University, wrote an intriguing book, *The Gospel According to the Beloved Disciple,* identifying that largely unidentified central figure of John's gospel as Mary Magdalene. Carse's book was a wondrously hypothetical concoction of what Magdalene's personal account of the event of Golgotha might have been—one that differed dramatically from any other written to this day and that neatly undermined nearly every tenet of classical Christian belief on the subject. I have read and reread Carse, and each time it has brought me to reexamine yet again my understanding of the entire issue—which is what mythical overlay is designed to do.)

I found less intriguing the subsequent wealth of books on the subject of Magdalene and Jesus that have generated a growing wave of public interest, with many an author riding that wave for all it's worth. Yet, as my editor Elaine pointed out to me, though I expressed my disdain and skepticism for such exploitation, all this recent ferment clearly illustrates my own contention that the greatness of mythical figures lies in their ability to stir up just such ferment within us. For those with ears to hear, this can lead us to abandon sterile notions and open again to the rich newness that these great figures endlessly bring about. (We can recall the apocryphal claim of Jesus: "I am always becoming as you have need of me to be.")

Surely, as Elaine also pointed out, building a new mythology on the love of these two memorable and divine characters could be a real step toward resanctifying sexual love itself, lifting it out of the tawdry muck and mire of current corporate-cultural degradations, with their flood of virtual realities that drag sex down to a gutter mind-set while linking it with ever more intense forms of violence. In typical cultural looping, this lucrative practice produces the very mind and market for more of such trash, feeding aggressively on the young adolescent in particular.

At any rate, as a result of mythological overlay, asking the real Jesus to stand up is futile, though he might stand quite tall in the divine imagination of those of us caught in his overall drama of soul, which is where the true power lies anyway. In my own case, at several pivotal crisis points in my adult life, this mythical image of Jesus formed within me since birth (as I mirrored my parents, who themselves mirrored theirs), concretized in my childhood, and was rejected in adolescence. But when I was around age thirty, Jesus made a dramatic resurgence in my mind's eye, which clearly shifted me into a less chaotic mode of mind and even, on occasion, indicated a particular decision I should make or actual direction I should take. Further, this mythical figure seemed to fill me with the strength to carry out these decisions. Regardless of which part of my psyche was talking to which, my experience showed me that Julian Jaynes's bicameral mind is always with us in some Laski Eureka! fashion—and that things aren't quite as simple as we might prefer.

SOCRATES AND JESUS

For me, the classical accounts of the crucifixion have always been roughly resonant with Plato's account of his mentor Socrates. This resonance is one I have long harbored, for like Jesus, Socrates found himself in a double bind, caught between dishonorably betraying his loyal following (which his proffered exile would have been) and honorably accepting his death. His state execution would, in effect, affirm and keep alive, at least in his disciples' memory, the core beliefs and principles he had espoused. It therefore may not have been so much Plato's depictions of Socrates' philosophy and character, as his description of that far more impressive willing acceptance of death that emblazoned Socrates' name in history. Had he died at a ripe old age, however nobly and cheerfully, surrounded by his weeping followers, and had his philosophy then been picked up and amplified by Plato, history's memory of him would probably have faded away—and, likewise, Plato's recollections and poetic inventions may have had less staying power as well.

"Let us now praise famous men and our fathers who begat us," reads Ecclesiastes, referring to those of whom we know and have some record. The quote goes on: "though some there be who have no memorial, and who have died as though they had never been." Socrates' memorial was Plato's account of his death, not necessarily his philosophy or his disciples' adulations. Needless to say, were it not for his crucifixion, the astonishing cosmological insights of Jesus may also have been lost to history.

Jesus's interpreters and chroniclers, which were legion in the decades after his death, gathered a large collection of fragments concerning him and copiously added to these. Constantly sorting out, collating, editing, and formatting freely, these chroniclers and interpreters filled in the holes of their ignorance with imagination and at times near-genius. They also filled in with classically appropriate sayings from and references to the equally mythical Hebrew scriptures. These ancient writings on which the Hebraic tradition was perpetuated were used ipso facto to give Jesus affirmation and substance, of sorts, while also saddling him with fatal baggage he didn't really need.

Julian Jaynes notes that during the axial period, the entire Middle East, as articulated in the Hebraic Judeo-Christian scriptures, was undergoing the historical-evolutionary shift of brain-function we have discussed here. Jaynes found parallels for his thesis in both these biblical scriptures, old and new, and in the mythical history of the entire Middle

East of the axial period. As we shall see, Rudolf Steiner's *Approaching the Mystery of Golgotha* also throws light on this theory.

Steiner's insight into what may have been going on in this axial period can be found in his rather elaborate structures of thought concerning a cosmic struggle between what he calls the Ahrimanic and Luciferian forces. Ahriman was an ancient god of darkness, while the name Lucifer comes from a term for light, lucidity, enlightenment, or clarity of thought and was the name chosen in mythology to describe the angel sent from heaven bearing light to the world. (Of course, "light of the world" was a mythological appellation attached to Jesus.) Steiner recognized that on that grizzly hill, with its "sticks and yardarms" on which the Son of Man was hanged, nature's evolutionary enigma, long in process of seeking a solution through trial and error, was moving toward a possible conclusion: The intelligence of the heart, that Darwin 2 higher force of benevolence and love, had risen to the occasion through this sacrificial figure who manifested that intelligence—possibly for the first time in history as we know it—only to get strung up for his trouble. This heart solution presented by Jesus would have brought about the complete dissolution of that age-long accumulation of field effect we have referred to as dominator culture—and culture as a self-surviving psychic entity was not about to let this happen . . . not then as not now.

Steiner's Ahrimanic and Luciferian figures are symbolic of the evolutionary enigma brought on by the ancient amygdala—and provide a way of bypassing that roadblock. Ever ready by its nature to throw us humans back into the ancient reptilian survival-defensive modes of brain, the amygdala is where the replication of culture, generation after generation, really takes place. This was the powerful vicious cycle nature attempted to break in the event at Golgotha.

So Jesus arose at the height of the chaotic axial period, and, as happens again and again with those great beings who are "lifted up," when all was said and done, he was not so much a victim of the Romans and that cross as he was of his chief chronicler and mythologizer, Paul. Jesus in fact disappeared in the concoctions of Paul, who, once on the scene, turned to his own advantage (as is the way of many followers) all the patched up scrapbooks that gave Jesus some background—despite the fact that, though Paul's ego needed this, the event at Golgotha did not.

As is so often the case, this extraordinary and giant character Paul himself became the nucleus of extensive mythological overlay that was destined to eventually overshadow the original figure he exploited. As *The Biology of Transcendence* discusses at length, if we look closely at this Pauline phenomenon, we can see some of the roots of the mess we are in today and why we have faced trouble for two millennia, instead of living in the peace and harmony of life that that prince of the heart tried, and is still trying, to bring.

15

EUREKA! MOMENTS AND CRACKS

God only Acts and Is, In existing beings or Men.
William Blake, *The Marriage of Heaven and Hell*

Two major Eureka! events can be proposed for this axial period, the first being Jesus's metanoia, which led him to Golgotha, and the second stemming from events following that Golgotha itself. An age-long quest of the Hebrew people for a God of love had produced the answer that could have broken through the shell of their cultural egg—and thus ours, because it would have initiated a species-wide shift of consciousness. Instead, the second Eureka! quickly sealed that crack the first had initiated.

Cracks in the cosmic egg are generally sealed almost immediately by the egg itself, a point I brought out in my first book, *Crack in the Cosmic Egg,* a half century ago. The crack that was Jesus's experience and the subsequent translation, concretizing for all to see, could have given us not only a new face of God but of ourselves and a new understanding of the way this wondrous world works. This crack should have been a light bursting into culture's cave of mind, but this would have erased that cultural effect and thus was countered in short order.

Culture as a field force hasn't changed since the event at Golgotha, other than undergoing an exponential increase in negativity. Dark dimensions line up the same today on a global scale—perhaps pivoting around that same Middle East area. History can be cyclic, though nature may have to give up the game this time around, for we seem intent on destroying the playing field itself.

THE SEALING OF THE CRACK

Consider how Gordon Gould's Eureka! represented a summation of the field of optical physics up to that point in time and brought into being a new aspect of creation that did not exist before. Around this new aspect a new physical field effect formed and has been building since, and it, too, may reach its own saturation point until another Eureka! occurs, and on and on it goes in a cyclic, evolutionary process.

In the same way Jesus's Eureka! summarized and was the high point of a long-building Hebrew/Jewish history. In this Eureka! summation he also brought in a new aspect of creation that did not exist before, and out of this a new field effect could have been built, one from which we as a species could have evolved, a reciprocation between mind and the fields of potential mind.

French historian and literary scholar René Girard wrote on the value the Hebrew scriptures hold in our history. Yet perhaps this evaluation is valid insofar as we place the Hebrew scriptures in the same category as we might Darwin 1, with Jesus ushering in Darwin 2. The Darwin 2 of Jesus's Eureka! should have automatically brought to summation and completion the Darwin 1 of the Hebrew scriptures, thus giving these scriptures more value without replicating their travesty and gore.

As we have seen in part 2, in child development each new stage arises out of the foundation of its preceding stage and incorporates that predecessor into its own operations, thus lifting or transforming the older or lower into the general nature of the newer or higher. This transformed synthesis is, in turn, incorporated into higher cycles as they emerge. These strange loops are like those eternally "spiraling gyres" found in poet William Butler Yeats's rich imagery.

In the same way that Gould's Eureka! (laser discovery) lifted the field of optical physics, Jesus's Eureka! should have lifted the field from which it arose (ancient Israel) into a new order of functioning that, sooner or later, would have spread worldwide. This lifting up, however, though attempted again and again throughout our history, has never yet been fully achieved. Instead, we find repeated the devolutionary breakdown of evolution's loop. If we consider, as Jaynes did, the long centuries involved in the transition from the bicameral mind to an individual mind, the translation of Jesus's Eureka!, which involved this very transition, almost surely would have taken generations to complete, had it been allowed. It was Paul's intervention, however, that broke the loop, and humanity reverted back to a previous state of mind without knowing anything had been lost.

Thus Paul's Christology usurped Jesus's God in the heart and reinstated old Jehovah himself, resurrected out of the Hebrew scriptures. The God toward which Jesus pointed is found only in the heart, that creative force that rains on just and unjust equally, without judgment. Paul, however, presented Jehovah in new garb, a God who claimed to be the epitome of love, but who raved away as always about laws, demands for sacrifice, foment of wars of retribution, and revenge without end. In effect, the summation of Hebraic history through Jesus and his heart was canceled by Paul and his Christ, and history's failings were reinstated pretty much intact.

It was Paul the Apostle, then, the most quoted authority in Christendom and its loudest voice, who sealed the crack that Jesus created in the cosmic egg. Paul set up a cultural rebound, eventually establishing the most powerful social-behavioral control yet achieved. He conceived the core of this countering drama through his own typical Eureka! revelation, though its effect was quite opposite that of Jesus. (Again, there is no sacrosanct aspect of the Eureka! effect, just as there is no morality. It occurs regardless of positive or negative implications. Indeed, culture pulls the strings in most Eureka! cases—and there is scant morality there.) Other Eureka! moments surely broke through during that fervent time after Jesus's death, experienced by other passionate, liminal misfits. We can remember that the more intense a field activity, the greater the incidents of lightning strikes—and the more intense the countering actions. The governing Romans made the previous priestly reactions tame by comparison, and the roads were lined with crosses for decades.

While Christian history was largely Paul's invention, he established the fame of his own Eureka! through it. The baptism of Jesus in the Jordan is a workable enough myth to account for Jesus's Eureka! and metanoia, but Paul's own Eureka!, which he advertised so long and loud, may have been authentic. Ironically though, the fuss he made scrabbling for authenticity canceled to a large extent that field effect of love that preceded him and gave him the impetus for his whole play, his self-emulating campaign may have helped bring down the temple priesthood he opposed. At the same time, however, Paul's ploy laid the foundation for two subsequent millennia of ongoing violence and a priesthood that still reigns today.

PAUL'S EUREKA! AND ITS TRANSLATION

For several years, scholars surmise, Paul brooded over his fundamental antipathy to this temple priesthood but also over his having joined those priests in their scramble to wipe out an irritating mob of protesters and promoters of yet another religious wave sweeping Israel as the story of Jesus's resurrection attracted more and more adherents.

The archetypal source of Paul's revelation lay in the Hebraic tradition in which the people struggled to survive in the troubled waters of the Middle East, with its panoply of gods to which the Hebrews added. As a people, they were often made subject by invaders, which they invaded in turn. The long imaginative history of the Hebraic creation of a God with whom they conversed and their eventual lament over their God having gone silent gave the foundation for the gathering cloud that produced the Eureka! answers of both Paul and Jesus, however light and dark these answers might be.

As part of the buildup to Paul's Eureka!, that archetypal image of father sacrificing son, given to the Hebrews through Abraham's famous story, had etched its way into the whole Middle East mind-set through centuries of cultural practice. The scene was then set for Paul's arrival—a long, slow buildup of the dark forces of an angry God and sacrificed human victims on the one hand, a cloud of negative energy, and at its opposite, a people's longing for a God of love and benevolence. Indeed, this longing of the heart was sung throughout the Hebrew scriptures in passages as divine as language can express right alongside the demonic ravings of Jehovah, at best a God in process. Thus the Hebrew scriptures were a graphic portrait of both the divided, contradictory creature we are and the nature of contradictory gods we create in our constant war within ourselves and without.

As suggested before, the expression of the positive longing for a God of love led to the event at Golgotha, from which, in turn, arose quite opposite events that built up from the archetypal cultural cloud forming after Jesus's death and that were unconsciously expressed by Paul himself as he wrestled with his conscience over the event at Golgotha. Thus we have the story of Paul (or Saul, as he's referred to pre-Eureka!), who, in confusion of mind, plagued as Odysseus was by doubt and uncertainty, wanders his own sleep-walking way toward Damascus. Beneath the surface lie his long-festering antipathies, conflicts, longings, and loves, and his mind is a temporary blank. At this point, in true Laski style, he was hit by that thunderbolt out of the blue, as Jesus had been before him. We "sow" winds of one hue and "reap" whirlwinds of all colors and dimensions.

After the Damascus event, which symbolically struck Saul blind with the brilliance of his own vision, the reborn Paul had to translate his Eureka! into the common language of his background. Scholars propose that several years may have elapsed between the Eureka! on the Damascus road and Paul's final written translations of it. In fact, he never finished his writing—it just kept coming—but he had hit on a synthesis that seemed to reverse the cultural power Paul opposed. While his effort would instead eventually support that culture's field effect, once he had his system worked out, Paul moved into action as one possessed.

His Eureka! and its translation were both extensions of an imaginative creation that reflected or emerged out of that central archetype of his people's mythical heritage. He proposed that the long-silent and absent God around which temple life and Jewish Law had centered had, on behalf of humanity itself, sent his own son to earth as a sacrifice to himself on that earth's behalf. On completion of the bloody deed, God drew his son back to himself. This follows a peculiar, convoluted logic—illogic, really—an aspect of which will actually play a part in the ensuing drama.

In Paul's translation, this new son of God, in a convoluted reversal of Abraham's mind-set, was sent as substitute for that lamb or ram of long ago that apparently didn't quite finish the job. Indeed, Jesus is even referred to as the Lamb of God. Paul's was a truly revolutionary idea that, like all such ideas, inadvertently incorporated and eventually became that which it would have replaced or overthrown. Thus in time, Paul's translation gave us yet another batch of priests feeding on the social body.

The irony of Paul's story is his recognition that we, as a semimad people, are dealing with not just petty political-military-monetary issues of the day but also "principalities and powers." Yet Paul seems to have played into the hands of the very power he moved against and threatened, but such is the nature of revolutions, which always seem to reinstate the conditions around which the original revolt generated.

UPSTAGING JESUS

Thus arose Paul's Christology in which Jesus disappeared and a vaporous heavenly figure formed as the archetypal image of an ethereal otherworld. Toward this image and the afterlife of this otherworld all attention began to center. Like Moses in Jewish antiquity, Paul's Christ was also backed by the Law—the very Law Paul had opposed yet had become a chief exponent of in new dress. Paul's Law of love supposedly made obsolete all other laws (much as we fight wars that are to end wars).

Law itself, however, is, like culture and its wars, a primal error regardless of dress, setting in motion endless tangles of sorrow and conflict.

Paul's copious writings to put this new Law into effect spawned more and more laws, and his invention was picked up and filled in by many imitators. Various gospel injunctions were attributed to Paul and later preserved under his name in this New Covenant, and these Pauline configurations concretized, making Paul's own cosmology tangible and available to all in clear and at times beautiful and powerful written form. (It is traditionally said, interestingly, that Jesus himself, Paul's supposed model, eschewed writing.) Paul's new cosmology and its ever burgeoning laws concerned a God in the sky straight out of those familiar and paradoxical Hebrew scriptures—but one with whom humanity could reconnect if they but followed the injunctions and instructions of Paul himself or those of his multiple imitators who followed him. Whatever the intricately interwoven details, his Christ crowded Jesus out of the scene and off the stage. Taking the spotlight in a kind of identity theft, Paul's Christ became Christ-Jesus and took over history.

Thus the tangled web that is the hallmark of a collapsed bicameral mind was further entrenched by Jesus's appearance, rather than resolved by a truly new mind for humanity, which was Jesus's intent and is nature's destiny. A new mind describes Jesus's own Eureka!, and had he been heard and understood, he would have revolutionized that society giving birth to him and societies to come. He brought to summation the Hebraic history birthing him—he was their finest hour—only to eventually be used against them.

Rudolf Steiner's own mythological construction of the mystery of Golgotha recognized that on that hill with its "sticks and yardarms," an ancient evolutionary enigma of nature, long in the process of seeking for solution, moved toward a possible conclusion. The solution came through that most positive force field of evolution itself, love and altruism, which had built to a saturation point by the longing of the heart of that particular society and all humanity. The intelligence of the heart then moved as love to experience this new self that alone could bale this floundering species out of its dilemma. Life was trying to rise to the occasion through this figure who manifested that intelligence outwardly for the first time, but whose effort was undone.

Love may not have needed such manifestation in previous ages, when the heart-brain-mind connection may have been intact and functioning. Love should be simply human life living itself without conflict. After all, if Darwin 2 was correct, love is a principal evolutionary force bringing us about, and if Jesus was correct, God is that love and life itself, and if Plotinus was right, love is life expressing itself outwardly as a universe, making our individual being what the whole show is about. Somewhere along the way this knowing was lost and culture as a process took over, but long millennia were involved and we have no record of what might have transpired. Love in its true expression needs no written records or stone monuments, only the ever-present moment of its fulfillment.

For now, we can consider that a reassertion of love through the event at Golgotha should have brought about the complete dissolution of the age-old accumulation of culture's negative field effect. We could not remain enculturated if caught up in that image on the cross, nor could we enculturate our children. But culture as a psychic entity was not about to let this happen. For one thing, we can't get caught-up by that figure on the cross if the cross is empty and we are caught up instead in a cloud nine mirage of Paul's Christ that we can manipulate endlessly in theological debate and rationalization and use to justify any action of revenge and "justice." "Christ's Holy War," a famous and recent Catholic father (a Cardinal!) called our diabolical Vietnam holocaust. "Onward Christian Soldiers!" cries our twenty-first century savior of Christ's democracy. But whom, we might ask, would Jesus bomb?

Paul's creation deflected away from Jesus and toward his Christ the longing of his people for cultural deliverance. This compromised and severely hampered nature's attempt to reestablish or resurrect the orbito-frontal loop and heart connection, the only way to overcome a dysfunctional bicameral system. Through his metanoia, Jesus was a biological corrective for us, the logic of life trying to straighten out its system. His cosmology grew out of and reflected creation and its evolution.

Imperative to our makeup is our need for a model of a new possibility, and this possibility presented itself through that exemplar on the cross. Yet the climax and conclusion of the long quest that led to the axial period itself was looped through Paul and the reptilian defensive mode underlying culture. Culture's deadly dictum of "Do this or else!" took on cosmic, "eternal" and terrifying significance through Paul's God of justice and his right-hand executor, Christ.

Paul's Christ became not only the Law-Giver but the Judge of the misbegotten humankind that had lost heart and wholeness of mind. Michelangelo's depiction of this Christ as Judge in the Vatican's Sistine Chapel magnificently depicts the ambiguity of this final synthesis and the tragedy of our species. If we study that face, we see the paradoxical mixture of the infinite patience and compassion of a God of love mixed with a cold indifference to the endless suffering supposedly being inflicted on half of humanity, no doubt for its own good.

And what has lost out? The Great Mother, nature's wisdom and intelligence that Rudolf Steiner called the Sophia, She, the Soul and heart of the world, the Vedic God's bride-to-be rescued from darkness, that which Robert Sardello recognizes as the creator and source of that wisdom of the world that comes through the heart. What has lost out is this evolutionary spirit of the future unfolding as the present moment, which Jesus so perfectly exemplified. Paul's all too masculine dominator force reinstated and strengthened that very destructive culture that Jesus had moved beyond and would have so moved us.

We can recall again from part 2 that question of nature at each human conception and the developmental stages following: Can we go for higher intelligence this time or do we have to defend ourselves again? On the greatest level yet in our sad history, this question had been asked, and for the first time it was answered positively through Jesus, who was birthed by his culture and society just for this. Here at this critical evolutionary moment added to through the ages by humankind's longing, this Sophia had loomed over her creation, searching for that resonant mind that could receive and translate her answer to humankind's quest. In new guise, Sophia's resonant attempt to unite our split and fragmented mind came out of the desert of outcasts to make his pitch, laying bare our true identity for all to see—and almost immediately he was countered by the negative once again, from within the same social structure.

The Jewish people lost out in this struggle of evolutionary-devolutionary forces, for this answer given, which they had set in motion, was used against them—and by one of their former upholders of the Torah. The impact of Paul's reversal of evolution's answer led to one of the great ironies of history: Jesus, making the sacrifice he did for the people he loved, bringing their long history of struggle to what should have been a triumphant conclusion and breakthrough, was used against those very people by the Christians that replaced him. That anti-Semitism could be born out of that cross meant to be the Semite's deliverance surpasses all understanding. Yet this particular political ploy was but one of a vast demonic web being woven as tightly today as it ever has been. Such are the pitfalls of a stochastic creation.

Religion, so focused on God and cloud nine and afterlife, tends to ignore life as a gift given in this moment, which, to be accepted, must be lived fully in this moment. Even more, culture tells us that first we must get our cow out of the ditch, bury our father, tend to this and that—all backed by that grim "Or else!" should we do otherwise. Culture tells the child he can't live in the joyful world of play but must instead spend those magical, precious years in grim preparation for that which never arrives. Treating this miraculous gift of life as only preparatory to some cloud nine fantasy is culture and religion's great seduction. But our life isn't a dress rehearsal for some vapid, abstract eternity. It is the big show itself, and our living earth is the place and our body our means. As Elizabeth Barrett Browning wrote:

Earth's crammed with heaven,
And every common bush afire with God;
And only he who sees takes off his shoes . . .

16

ORIGIN AND FIELD

Impulse from below and forward movement from above
Who really knows? Who here can say?
When it was born and from where it came—this creation?
The gods are later than this world's creation,
Therefore who knows from where it came?
That out of which creation came,
Whether it held it together or did not,
He who sees it in the highest heaven,
Only He knows—or perhaps even He does not know!

Rig Veda

The Rig Veda sings of the ever-present Origin and its endless gates opening to field effect, Laski's Eureka! phenomenon, creator and created giving rise to each other in reciprocal strange loops, Darwin 1 as impulse from below, Darwin 2 as forward movement from above, Sardello's love and the world, Eisenstein's separation-reunion, steady-state creation, big bangs, what have you. Who really knows? Who here can say from where this creation came?

Religion, however, for those buying into it, is an answer that shuts down the Rig Veda's question and the dialogue toward which it points, a communication between impulse and forward movement, the fundamental strange loop from which our reality springs. That love and altruism are the forward moving forces of evolution is a top-down function brought to full conscious awareness through that bottom-up event leading to Golgotha. Religion projects onto cloud nine an answer that, were it the case, would bring closure to the loop—an impossibility.

The fundamentalism arising on every hand today is brought about by the growing awareness of the death of religion that is underway—and religion as a concept and cultural force is, as we learned at the beginning of this book, a fundamental plank in the platform of culture's ideation. The extreme behaviors of fundamentalists have become a profound threat, and to argue here on behalf of an authentic foundation for the mythically overlaid and well-tarnished image of Jesus at such a time as ours seems counter to reason. Yet this book has argued on behalf of the cosmology of mind in creation that he presented, which was such a powerful countercultural force that it brought on the event of Golgotha through which it was effectively silenced. Culture as a destructive force has grown steadily ever since and its schism of mind and creation now reaches toward a final and ultimately disastrous climax.

It is telling that today, in almost any setting or gathering, we can openly discuss the philosophy of Socrates, for instance, and feel at ease and appropriate, but if anyone should mention the philosophy of Jesus, an embarrassed silence falls. Such is the price we pay to Christendom, bringing to mind Robert Wolff's comment concerning the Senoi in Malaysia: we have no idea what we have lost.

ALLOWING THE MIND-HEART DIALOGUE

The time-worn episodes of a hypothetical axial period some two millennia ago graphically portray the issue of an evolutionary setback in our evolving brain that was coming to a nexus at that time. As we have seen, Julian Jaynes assumed the movement of mind from bicameral to individual was largely established back then except in isolated pockets about the world. Yet it seems that the evolutionary shift Jaynes explored did not hold and expand or, the setback not resolved, was certainly not completed, for a split of mind has plagued us for two millennia up to this day. We are again, as a species, approaching a critical bicameral breakdown, the issues that moved in that long-ago day now massively mounting, perhaps toward a grim closure in our own.

This book has followed Piaget's observation that mind is an emergent property of the brain, able at maturity to turn around and operate on the very structure that had produced it. Steiner saw this in a more expansive light and spoke of allowing the heart to teach us a new way to think in order for the heart to discover its own next level of evolution. The metamorphosis of thinking that the heart could then bring about would, in turn, give our heart its necessary reciprocating element for bringing a new creation into being. So a system, the heart, produces a product, the human mind, by which that heart system can then go beyond itself. But culture brought a schism between mind and heart, one that has derailed evolution.

Significant is Steiner's use of the word *allow.* This heart-mind loop is not an action we take, but a move we *allow,* precisely as for the reception of a Eureka! experience. George Fox waited for his openings of spirit, directives telling him which alternative directions to choose; Socrates spoke of his daemon directing him; and Jesus spoke of his indwelling spirit of wholeness lighting his way—all clear descriptions of the mind-heart dialogue. The allowing that is required involves both aspects of the strange loop: Each can only be by allowing the other. Religion attempts to stop this loop and freeze it into a projection of cloud nine or some object or force we can manipulate with our mind.

This movement between heart and brain is actually below our awareness, or mind, making necessary a reception-translation of such heart-brain products, which in turn requires a looping back yet again to bring a new possibility into our full, conscious awareness. Only through such reciprocal looping, through neural systems compatible with the materials given *by* those neural systems, could the answer be acted on with recognition and translation into a final realization taking place. If we grasp the importance of this strange-loop process, we will see the paucity and pettiness of both religions, scientism and fundamentalism.

Rudolf Steiner's challenge that we allow the heart to teach us a new way to think may be a challenge that has been given often and in different ways in the vast stretches of evolutionary time—analogous, perhaps, to the appearance of the prefrontal cortex at different times. Without sufficient support to be maintained, both new brain and that higher expression of intelligence we call thinking in the heart may have appeared and then atrophied and disappeared time and again, as may be the case today.

It might be that love as human, giving rise to us some forty or fifty thousand years ago, functioned as designed for long ages after we appeared. Why should it not have? Long ages prepared for it. The prefrontal-heart connection may have run the show quite successfully until it broke down at some point. Stochastic, random chance is always part of the process. Postwar trauma or massive disaster can alter brain development in subsequent generations, and those generations would then be unable to detect the change in their nature, which would always be simply the "human condition," assumed to be our nature. Humankind may have suffered near-irreversible setbacks from its two world wars of the twentieth century—and since that time, more and more stable functions have been coming apart, with dysfunction running amok.

But the field effect of love and altruism that produced us is still right there (wherever there is), even if its product—humanity—strips its gears or loses its track. Always moving for our well-being and fulfillment, this nurturing intelligence of heart moves to correct the course of its evolving product, humanity, though its evolutionary processes may be slow in our timeframe. Darwin's evolutionary force of love may have accumulated over the ages as a field effect leading to Golgotha. The force involved was and is all there in our human brain, but yet not entirely of it. This force, nature's wisdom and intelligence, Sophia, She, the Great Mother, functions through the heart.

We can recall again Steiner's observation of the heart as center of a universe that the heart collects into itself and radiates out simultaneously, an observation underlying Robert Sardello's spiritual psychology and spelled out in Kashmir Shaivism in the tenth century and developed and explored in quantifiable scientific terms by the Institute of Heart-Math and neurocardiology in our own day. This wisdom is the spirit of the future unfolding as that present in which Jesus cried out exultantly: "Behold I make all things new," and "I am forever becoming what you have need of me to be." Creating the world anew, instant by instant, rather than following that deadly religious path that replicates our past and all its sorrows again and again, is always our choice.

The translation of Gordon Gould's laser Eureka! revelation was quite difficult because it involved something new under the sun. The concrete operations on nature that it required were complex, but once established, as Sheldrake points out, any new phenomenon tends to become easier and easier, just as the laser procedure eventually became commonplace. The same should have happened with Jesus and his translation of his Eureka!—it should finally have been commonplace, simply the way things are, giving us a new level to lift to and go beyond. Of course, the answer that Jesus pointed toward was no simpler on its broad social-cultural level than Gould's in his rule-bound optical physics. Jesus's social-cultural answer could only be lived out, expressed through relationship. Gould's invention, on the other hand, depended on relating anew physical phenomena, which is easy compared to relating anew human emotions and reactions, to reversing the instinctual reflexes of the amygdala and reestablishing the bonds of heart.

As Lao Tzu had found long before, language is a poor vehicle to carry a true shift of mind. While Jesus wisely eschewed writing (which tends to freeze into dogma), he mined a rich vein of metaphor, simile, parable, allusion, fable, and story that might tell his tale, these means being flexible and ever-adaptable to our changing needs. His message from beginning to end was a new cosmology, a biological, nonmythical, nontheological, anthropomorphic description of how our reality unfolds and what our role in it is. (*Anthropomorphic* means arising from and relating to humankind, not some ethereal abstraction. We can note that Jesus never referred to himself as the Son of God, but rather as a Son of Man, as are we all.) That his translation was bent backward into a restatement of the very trap of culture from which he tried to lead us is simply the haphazard way of a stochastic creation.

In no way are we called on to reinvent the wheel he presented or try to follow in his footsteps, which would tie our future to the past. We are called upon only to allow the heart to move in response to such a model. The value of Gerald Feinberg's report on the firewalkers of Ceylon was the singular fact that such a phenomenon is possible—not that we should rush out and try to do the same, but that we should open our mind to that which lies outside our current knowledge, with its restraints and boundaries.

THE VASTNESS OF UNKNOWING AND THE NOW

The Rig Veda epigraph at the beginning of this chapter speaks of what came before the creation, before humanity and the myriad gods we have invented. Perhaps the ever-present Origin of which Jean Gebser speaks is that Vastness that lies beyond even the Ground of Being. Our link with both Ground of Being, from which creation springs, and the Vastness from which the Ground of Being springs, is through the heart. The heart alone can teach us this new way of thinking, which may be a first step in bridging the gap first between mind and the Ground, then between Ground and Vastness. This bridge into the unknown may be precisely the heart's evolutionary goal for our species.

Lao Tzu claimed there is no reciprocation with this Vastness. It is not a two-way street. Perhaps a two-way thoroughfare hasn't been brought about through evolution—yet. Meister Eckhart said that to enter that cloud of unknowing, all names, identities, and knowledge must be left outside; thus, in experiencing it and then trying to express it, we can only lapse into silence.

As do many people, I speak of the Vastness as absolutely other, a phenomenon at a radical discontinuity with brain-mind and its concepts, precepts, and percepts. What little I can say of this is from my own limited experience. At around age forty, as I explained briefly in

The Biology of Transcendence, I underwent a temporary absorption into a radical discontinuity with anything and everything I had known. Brief and fleeting as it probably was (though the event seemed timeless and I have no idea what a clock might have registered), the experience nearly ruined me. I wasn't worth much after that for quite some time. Of course, it would be both futile and foolish to attempt to describe the state or content of the experience, even to myself.

As suggested in William James's epic *The Varieties of Religious Experience,* I suspect that many people have undergone some aspect of this fusion with the Origin or Vastness but couldn't conceptually cope with it any better than I did. Perhaps, then, people automatically erase such experiences from mind to maintain their integrity. Something so absolutely other to anything comprehensible may be simply screened from mind or memory lest our ideation be dismantled. Yet the experience is so overwhelming that our subsequent loss of it brings an intensification of the heart's longing that it is nearly too intense to bear. Once having had such an experience, an individual can never forget its impact or forget the fact that this Vastness is always right here.

This Vastness may have been Eckhart's reference in his prayer, "Oh God! Deliver me from God." Nothing else he could ever experience would measure up to what he had experienced in that great unknowing, which can leave a person feeling bewildered, stranded, and estranged from the cultural world, never again to be comfortable in it.

Bernadette Roberts describes being caught up in bodily form and for weeks breathing a divine air, surely a fitting term for the Ground of Being as it touches on the Vastness. Coming back into her everyday world was horrible for her, she reports, with the contrast being too much to reckon. Surely something equally powerful if not more so had been experienced by that great being strung up on Golgotha two millennia ago—something so powerful that it locked into our collective conscious or unconscious and became itself a nucleus of constant overlay. This may explain why so many of us have intuitively gravitated toward that symbolic, mythical figure and feel such resonance with him, even as we may seriously reject the dark religion of Paul's Christology that has shrouded that light.

Perhaps Jesus's willing acceptance of his fate came out of the bewildering explosion of love we feel when seized by that absolutely other. Whatever the details, something of Jesus's state cemented that Golgotha event into our species psyche, and from that moment, that state was the target of all our spiritually inclined experiences and projections. Mythological overlay on this symbolic figure of love greater than self has acted ever since as a gravitational point for our worldwide longing for love. Whether or not we have ever heard of Jesus, any such longing of heart feeds into that same center. George Leonard speaks eloquently of his sixteenth year and the longing that he couldn't express but that was so intense he felt it could never be assuaged. A similar longing, however fleeting for some, exists in every teenager and perhaps in all humanity. All such experience is resonant with that target and would tend to gravitate toward it and thereby strengthen it. Similarly, Plotinus likens the cosmos to an expression of love searching for its own reflection. (He also reported absorption into that Origin seven times in his life. Lucky man.) Our search for and longing for love is at the same time a journey into God that could never end, for love as a force always lies beyond itself, like the horizon of an ever-expanding universe.

The event at Golgotha acts not only as a focal point but as a mediator of sorts between ourselves and that Ground of Being that can link us to the Vastness, concretizing an otherwise abstract concept or possibility and thereby making real the present moment in which it all happens. Nothing in Jesus's original statements or actions indicates that an afterlife was what he or life was about. As he pointed out, God was a God of the living, not the dead. God is life itself, with its underlying intelligence of creativity. Jesus pointed to that which opens us to this life today, in this moment. What we were to give up for greater life was our quaking subservience to culture and its dominator "Or else!," although the surrendering of such an imprint seems to our enculturated mind to be a collapse into chaos or apparent death. The world we should "hate" was the world of culture—not the victims of and within it.

Jesus's aim was to show us how this very moment contains all possibility—not just what is needed, but the whole shebang. Today is the day, this is the hour. He cautioned against making oaths of any sort, any form of contracts, legal agreements, and so on (even marriage) because they tie up our future in that we lose the ever-present spirit of wholeness and the moment. Making some great decision, no matter what or how noble the impetus, can stop us in our tracks. To decide is to cut off alternatives, which then locks us into a past decision that only closes off the present moment. Consider everything provisional; the event of the moment provides the future flowing into it, to which we can only open and allow. If we grasp this, we can see the Jesus event in its own light and we may also glimpse the full dimension of culture's dark scientism and religion.

This same caution to stay open to the present holds true in spite of any and all the noble principles on which we stand. With a stalwart stubbornness, we take pride in our principles and think ourselves virtuous holding them, and this noble virtue simply predetermines our actions, replicating our past and closing us to this moment's open spirit of newness. In the resulting state of isolation from the heart we truly have no provision—we can't see ahead, so we grope blindly.

Everything in Jesus's ontology-cosmology was designed to open us to that future flowing into the present, which is heart's presence, allowing the spirit of wholeness to unite our fragmented life into that single vision that can keep us open, regardless of the world of folly out there. No matter, then, that dark clouds rise on every hand today, as always. Sophia still offers her answer, centered in our heart as always, closer than our very breath, ready to breathe us anew each moment.

Now, star flake frozen on the windowpane
All of a winter night, the open hearth
Blazing beyond Andromeda, the sea-
Anemone and the downwind seed, O moment
Hastening, halting in a clockwise dust,
The time in all the hospitals is now,
Under the arc-lights where the sentry walks

His lonely wall it never moves from now,
The crying in the cell is also now,
And now is quiet in the tomb as now
Explodes inside the sun, and it is now
In the saddle of space, where argosies of dust
Sail outward blazing, and the mind of God,
The flash across the gap of being, thinks
In the instant absence of forever: now.

HOWARD NEMEROV, "MOMENT"

BIBLIOGRAPHY

Acuff, Daniel S., and Robert H. Reiher. *Kidnapped: How Irresponsible Marketers Are Stealing the Minds of Your Children.* Chicago: Dearborn, 2005.

Armour, J. Andrew. *Neurocardiology: Anatomical and Functional Principles.* Boulder, Calif.: Institute of HeartMath Publications, 2003.

Armstrong, Allison, and Charles Casement. *The Child and the Machine: Why Computers May Put Our Children's Educations at Risk.* Toronto: Key Porter Books, 1998.

Austin, James H. *Zen and the Brain.* Cambridge, Mass.: Massachusetts Institute of Technology Press, 1998.

Axness, Marcy. "The Trauma of Separation: Some Socio-cultural Perspectives on Motherloss." From Ph.D. dissertation for Union Institute and University Graduate College, 2003.

———. "Abortion and the Collective Soul: Questions of Autonomy, Responsibility, and Meaning." From Ph.D. dissertation for Union Institute and University Graduate College, 2003.

Bailie, Gil. *Violence Unveiled: Humanity at the Crossroads.* New York: CrossRoad Publishing, 1995.

Bateson, Gregory. *Mind and Nature: A Necessary Unity.* New York: E. P. Dutton, 1979.

Blake, William. *Selected Poetry and Prose.* Edited with an introduction by Northrop Frye. New York: The Modern Library, Random House, 1953.

Bohm, David. *Causality and Chance in Modern Physics*. London: Routledge and Kegan Paul, 1957.

———. *Wholeness and the Implicate Order*. London: Routledge and Kegan Paul, 1980.

Buhner, Stephen Harrod. *The Secret Teaching of Plants: The Intelligence of the Heart in the Direct Perception of Nature*. Rochester, Vt.: Bear and Co., 2004.

Caplan, Mariana. *Untouched: The Need for Genuine Affection in an Impersonal World*. Prescott, Ariz.: Hohm Press, 1995.

Cobb, Jennifer. *Cybergrace: The Search for God in the Digital World*. New York: Crown, 1998.

Damasio, Antonio R. *Descartes Error: Emotion, Reason, and the Human Brain*. New York: Avon Books, 1994.

Deardorff, Daniel. *The Other Within: The Genius of Deformity in Myth, Culture, and Psyche*. Ashland, Ore.: White Cloud Press, 2004.

de Caussade, Jean Pierre. *The Sacrament of the Present Moment*. San Francisco: HarperCollins, 1982.

DeMarco, Frank. *Muddy Tracks: Exploring an Unsuspected Reality*. Charlottesville, Va.: Hampton Roads, 2001.

Denton, Michael. *Nature's Destiny*. Old Tappan, N.J.: Free Press, 2002.

Eisler, Riane. *The Chalice and the Blade: Our History, Our Future*. San Francisco: HarperSanFrancisco, 1987.

Eliade, Mircea. *Cosmos and History: The Myth of the Eternal Return*. New York: Harper Torchbooks, 1959.

———. *Yoga: Immortality and Freedom*. New York: Pantheon, Bollingen Series, 1958.

Farcet, Gilles, ed. *Radical Awakening: Conversations with Stephen Jourdaine*. Carlsbad, Calif.: Inner Directions Foundation, 2001.

Forman, Robert K. C. *Meister Eckhart: Mystic as Theologian*. Rockport, Mass.: Element Books, 1991.

Gebser, Jean. *The Ever Present Origin*. Stuttgart: Deutsche Verlags-Anstalt, 1949.

Goldberg, Elkhonon. *The Executive Brain: Prefrontal Cortex and Civilized Mind*. New York: Oxford University Press, 2001.

Goswami, Amit. *A Quantum Explanation of Sheldrake's Morphic Resonance*. Eugene, Ore: University of Oregon Institute of Theoretical Science, Monograph, n.d.

Grossinger, Richard. *Embryos, Galaxies, and Sentient Beings: How the Universe Makes Life*. Berkeley, Calif.: North Atlantic Books, 2003.

Hannaford, Carla. *Smart Moves: Why Learning Is Not All in Your Head.* Salt Lake City: Great Rivers Press, 1995.

Harris, Sam. *The End of Faith: Religion, Terror, and the Future of Reason.* New York: W. W. Norton, 2004.

Harrison, Steven. *What's Next After Now?: Post Spirituality and the Creative Life*. Boulder, Colo.: Sentient Publications, 2005.

Hart, Tobin. *The Secret Spiritual World of Children.* Makawao, Maui, Hawaii: Inner Ocean, 2003.

Hartmann, Thom. *The Last Hours of Ancient Sunlight.* New York: Three Rivers Press, 1998

———. *The Prophets Way: A Guide to Living in the Now.* Rochester, Vt.: Park Street Press, 2004.

Hoffstadter, Douglas R. *Godel, Escher, Bach: An Eternal Golden Braid.* New York: Basic Books, 1979.

Jahn, Robert, and Brenda Dunne. *Margins of Reality: The Role of Consciousness in the Physical World.* New York: Harcourt Brace Jovanovich, 1979.

Jaynes, Julian. *The Origin of Consciousness in the Breakdown of the Bicameral Mind.* New York: Houghton Mifflin, 1976.

Laszlo, Ervin. *Science and the Akashic Field: An Integral Theory of Everything*. Rochester, Vt.: Inner Traditions, 2004.

Lehman, Christine. "Young Brains Don't Distinguish Real from Televised Violence." *Psychiatric News* 39, no. 15 (2004): 37.

Lipton, Bruce. *The Biology of Belief: Unleashing the Power of Consciousness, Matter, and Miracles*. Santa Rosa, Calif.: Mountain of Love/Elite Books, 2005.

Loye, David. *Darwin's Lost Theory of Love.* Lincoln, Neb.: iUniverse, 2000.

MacLean, Paul. "The Brain and Subjective Experience: Question of Multi-level Role of Resonance." *Journal of Mind and Behavior* 18, nos. 2 and 3 (Spring 1997): 247–68.

———. *The Triune Brain in Evolution.* New York: Plenum Press, 1990.

———. "Women: A More Balanced Brain?" *Zygon Journal of Religion and Science* 31, no. 3 (1996).

Malinowski, Bronislaw. *The Sexual Life of Savages in North-western Melanesia: An Ethnographic Account of Courtship, Marriage, and Family Life among the Natives of the Trobriand Islands.* Boston: Beacon Press, 1987.

Marinelli, Ralph, Branko Furst, Hoyte van der Zee, Andrew McGinn, and William Marinelli. "The Heart Is Not a Pump: A Refutation of the Pressure Propulsion Premise of Heart Function." *Frontier Perspectives* 5, no. 1 (1995).

Marshak, David. *The Common Vision: Parenting and Educating for Wholeness.* New York: Peter Lang, 1997.

McCraty, Rollin, Mike Atkinson, and Raymond Travor Bradley. "Electrophysiological Evidence of Intuition, Part 1: The Surprising Role of the Heart." *Journal of Alternative and Complementary Medicine* 10, no. 1 (2004): 133–43.

McCraty, Rollin, Raymond Travor Bradley, and Dana Tomasino. "The Social Heart: Energy Fields and Consciousness." Boulder Creek, Calif.: HeartMath Research Center Bulletin, Institute of Heartmath, 2004.

Mead, Margaret. *Coming of Age in Samoa: A Study of Adolescence and Sex in Primitive Societies.* New York: Penguin, 1965.

Montanaro, Sylvana Quattracchi. *Understanding the Human Being: The Importance of the First Three Years.* Mountain View, Calif.: Montessori, 1991.

Montessori, Maria. *The Absorbent Mind.* Madras, India: Kalakshetra Press, 1992.

Narby, Jeremy. *The Cosmic Serpent: DNA and the Origin of Knowledge.* New York: Tarcher/Putnam, 1998.

Nemerov, Howard. *The Selected Poetry.* Edited by Daniel Anderson. Athens, Ohio: University of Ohio Press, 2003.

Nichol, Lee, ed. *The Essential David Bohm.* New York: Routledge, 2003.

Nicolescue, Basarab. *Science, Meaning, and Evolution: The Cosmology of Jacob Boehme.* New York: Parabola Books, 1991.

Odent, Michel. *The Scientification of Love.* London: Free Association Books, 1999.

Pagels, Elaine. *The Gnostic Gospels.* New York: Random House, 1979.

———. *The Gnostic Paul.* Philadelphia: Trinity Press International, 1992.

———. *The Origin of Satan.* New York: Random House, 1995.

Prescott, James W. *The Origins of Human Love and Violence.* The Societal Impact. Monograph prepared by the Institute of Humanistic Science for the International Congress of the Association for Pre- and Peri-Natal Psychology and Health.

Roberts, Bernadette. *The Contemplative: Autobiography of the Early Years.* Copywrite manuscript available only through http://bernadettesfriends .blogspot.com/.

———. *What Is Self?: A Study of the Spiritual Journey in Terms of Consciousness.* Austin, Tex.: Goens, 1989.

Sardello, Robert. *Love and the World: Conscious Soul Practice.* Great Barrington, Mass.: Lindisfarne Books, 2001.

———. *Silence.* Benson, N.C.: Goldenstone Press, 2006.

Schore, Allan. *Affect Regulation and the Origin of the Self: The Neurobiology of Emotional Development.* Mahwah, N.J.: Lawrence Erlbaum Associates, 1994.

———. "The Experience-Dependent Maturation of a Regulatory System in the Orbital Prefrontal Cortex and the Origin of Developmental Psychopathology." *Development and Psychopathology* 8 (1996): 55–87.

Schurman, Reiner, trans. *Wandering Joy: Meister Eckhart's Mystical Philosophy.* Great Barrington, Mass.: Lindisfarne Books, 2001.

Segal, Suzanne. *Collision with the Infinite: Life beyond the Personal Self.* San Diego: Blue Dove Press, 1996.

Sheldrake, Rupert. *A New Science of Life.* Rochester, Vt.: Park Street Press, 1995.

———. *Dogs That Know When Their Masters Are Coming Home and Other Unexplained Powers of Animals.* New York: Three Rivers Press, 1999.

———. *The Sense of Being Stared At.* New York: Crown, 2003.

Tiller, William, Walter E. Dibble Jr., and Michael J. Kohane. "Exploring Robust Interactions Between Human Intention and Inanimate/Animate Systems." Monograph presented at "Toward a Science of Consciousness: Fundamental Approaches." May 25 to 28, 1999, United Nations University, Tokyo, Japan.

Velmans, Max. *Understanding Consciousness.* London: Routledge, 2000.

Williamson, G. Scott, and Innes Pearse. *Science, Synthesis and Sanity: An Enquiry into the Nature of Living.* Edinburgh: Scottish Academic Press, 1980.

Wilson, Frank. *The Hand: How Its Use Shapes the Brain, Language and Human Culture.* New York: Knopf, 1999.

INDEX

BOOKS OF RELATED INTEREST

The Biology of Transcendence
A Blueprint of the Human Spirit
by Joseph Chilton Pearce

The Crack in the Cosmic Egg
New Constructs of Mind and Reality
by Joseph Chilton Pearce
Foreword by Thom Hartmann

Spiritual Initiation and the Breakthrough of Consciousness
The Bond of Power
by Joseph Chilton Pearce

From Magical Child to Magical Teen
A Guide to Adolescent Development
by Joseph Chilton Pearce

Science and the Akashic Field
An Integral Theory of Everything
by Ervin Laszlo

Stalking the Wild Pendulum
On the Mechanics of Consciousness
by Itzhak Bentov

Original Wisdom
Stories of an Ancient Way of Knowing
by Robert Wolff

Radical Knowing
Understanding Consciousness through Relationship
by Christian de Quincey